IMAGES
of America

MINES OF CLEAR CREEK COUNTY

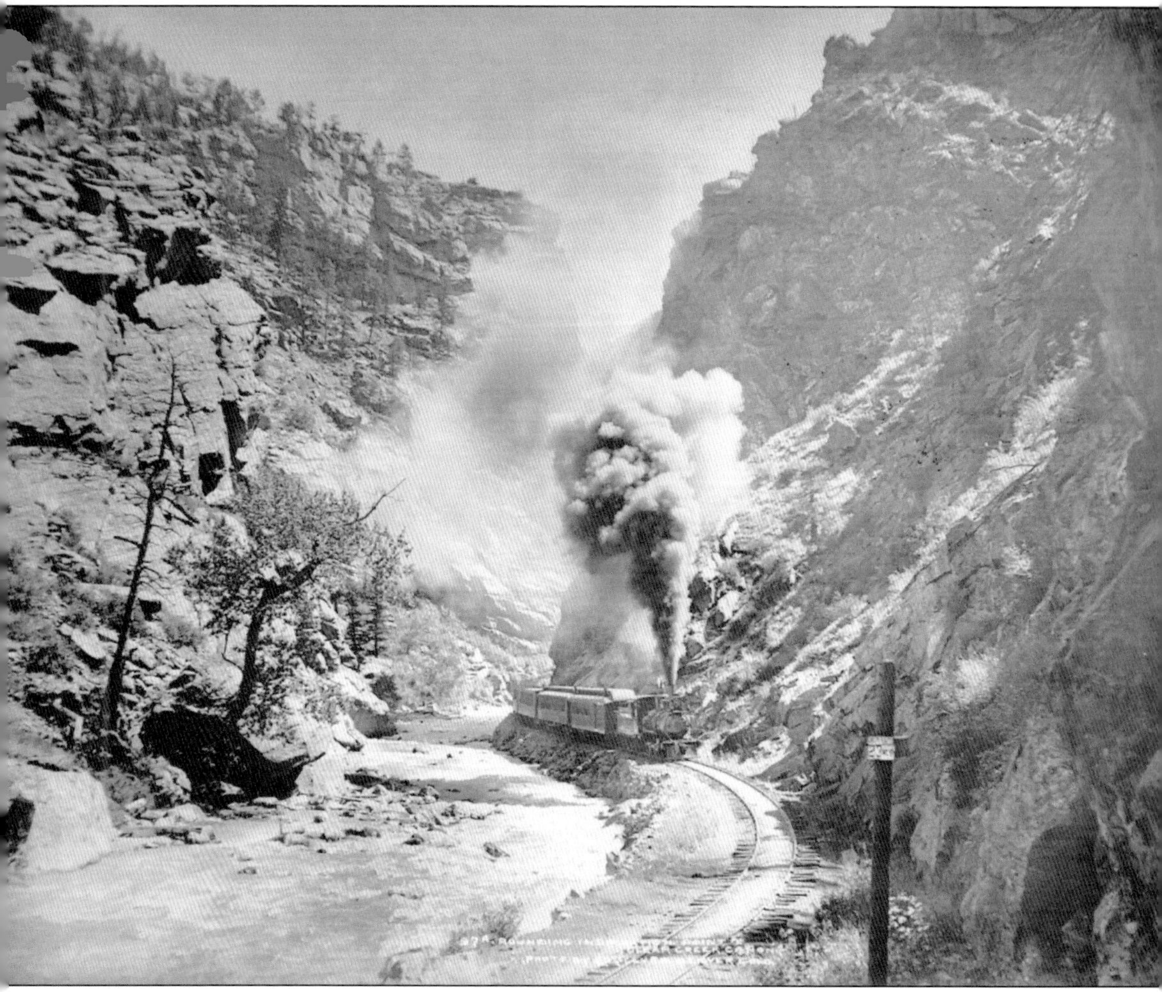

The Colorado Central Railroad train travels up Clear Creek Canyon to Clear Creek and Gilpin Counties. The arrival of the railroad brought new life to these communities. The tracks were first built to Black Hawk and Central City in 1872. A financial panic in 1873 caused a delay in getting the railroad to Idaho Springs and on up the Clear Creek Valley until 1877. The railroad made it easier to transport ores from the mines to the mills and smelters. (US Geological Survey.)

On the Cover: This Georgetown scene took place in front of Walsh's Saloon in the early 1890s. Ulrick Walsh operated this saloon from 1889 to 1915. Coors beer was brought to Georgetown from Golden. Adolph Coors began brewing and distributing beer to the 30 miners of Colorado in 1877. Transport was undoubtedly made easier by utilizing the railroad. Life in Georgetown in the 1880s was very prosperous, but things changed in 1893—there was a financial panic caused by the repeal of the Sherman Silver Purchase Act. The act, passed in 1890, did not authorize the free and unlimited coinage of silver but increased the amount of silver the government was required to purchase. (Denver Public Library.)

IMAGES
of America

MINES OF CLEAR CREEK COUNTY

Ben M. Dugan

ARCADIA
PUBLISHING

ISBN 978-1-4671-3034-9

Published by Arcadia Publishing
Charleston, South Carolina

Printed in the United States of America

Library of Congress Control Number: 2013945090

For all general information, please contact Arcadia Publishing:
Telephone 843-853-2070
Fax 843-853-0044
E-mail sales@arcadiapublishing.com
For customer service and orders:
Toll-Free 1-888-313-2665

Visit us on the Internet at www.arcadiapublishing.com

This book is dedicated to Ron Neely (late) and Cindy Neely; Don Weber; coordinator of the Georgetown Guides Christine Bradley; and to my wife, Pam, and son, Sam, for all the time away from home in service to this book.

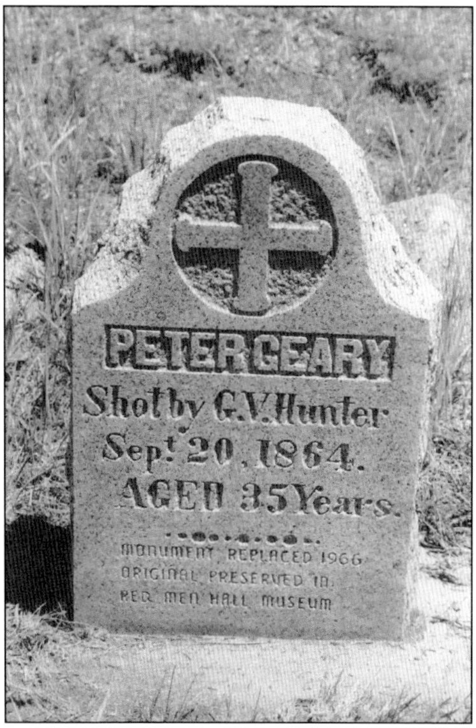

A small town comes of age. An early killing in the small town of Empire shocked local residents. Peter Geary, a peace-loving Irishman, had two dogs that gave G.V. Hunter's oxen some trouble. Hunter shot one of the dogs and then Geary. The stone marker was the first of subsequent stones to be placed in the new cemetery on the bluff in the southeast corner of town. Prior to the stone marker, only rocks had been piled on graves to protect them.

Contents

Acknowledgments

This book is an effort that is a lifetime in the making. Growing up in the small mining town of Empire, Colorado, and attending elementary school in the mining town of Georgetown five miles to the south certainly influenced the author's formative years. Georgetown Society was formed in 1970 as a historical preservation entity for the rich and unique history of Georgetown. The society sponsored an educational program, the Georgetown Guides, at the local elementary school. In the late 1970s, many children were involved with this program and now have their own unique perspectives on what the history teaching did for them. These students include the following: Mike Westjohn, Shannon Cardinal, Aileen Martinez, Mark Babeon, Ryan Tofflemire, Barb Fiveash, Angie Pennell (Morgan), Richard Dude Garcia, Dena Bandazian (Nishek), Ann Wale (Huml), Cindy Jeffrey, Tammy Hufford (Woolery), Dan Bane, Brenda Barela, Belinda Barela, Nick Goloskewitsch, Craig Abrahamson, Ben Dugan, Troy Cardinal, Jodi Buckley, Tina Brethauer, Mary Rutherford (Albertoni), and Lisa Hammond (Watson). Teachers Nancy Dudek, Dean Halbfoster, school principal Don Weber, Robert Gibbs, and society director Clay Hutchinson were also involved in the program. One of the projects these students completed was a full-scale block-by-block diorama of 1890 Georgetown. In 1979, they worked in small groups, with each building a city block of out of cardboard.

Most of these students also went to high school in Idaho Springs, growing up in and around that mining locality on a daily basis, taking a school bus going up Fall River, and passing the Lucania Tunnel and other mines along the way. Many passed the Alice Glory Hole on their way up to Loch Lommond to go fishing. There are remnants of hundreds of old mines everywhere in Clear Creek County. There are over 4,000 patented mining claims in Clear Creek County; a patented mining claim refers to a claim needing annual improvements in order to maintain possession and ownership per the 1872 mining law.

Additional appreciation as well as thanks are directed toward Wallie Robinson, Coi-Drummond Gehrig (at Denver Public Library), Jennifer Dibbern (at History Colorado), Colorado School of Mines information, and photographs provided by Ann Wale (Huml) and husband, Kirk Huml. Arcadia acquisitions editor Stacia Bannerman, and Robert Putfark, Jim Prochaska, Robert Mangione, Duane White, Mark Alexis, and Daryl Boone of the Glenbrook Gallery (Redmen hall) of Empire also contributed to the final effort. A special thank-you goes to Gary and Sally St. Clair of the Peck House, David Hays and Charles Jacoby at University of Colorado Archives, and to Mel and Helen Vogtman for taking care of my little son, Sam, while I was off doing research. Lynn Schimmelfeder also provided a great many photographs of Clear Creek County mines in the 1970s. Many of these old mines no longer exist. Her photographs of the Alice Glory Hole are some of the few that remain in existence and are shown in the Fall River section. This book was also made possible with a generous donation from the Clear Creek County Historical Society. Images used are credited to the Denver Public Library (DPL), University of Colorado at Boulder, Lachlan McLean Collection (UCB), Library of Congress (LOC), US Geological Survey (USGS), or Clear Creek County Historical Society (CCHS), or are from the author's private collection.

Introduction

The cowards never started. The weak died on the way. Only the strong arrived; they were the pioneers

—Kit Carson (attributed)

The financial panic of 1857, caused, in part, by the sinking of the SS *Central America* returning from California overloaded with gold, upset the balance and health of the nation's economy. The event shook the confidence of the public. Warehouses became overstocked, employment was at a premium, and most families' savings were completely wiped out. There were tough times nationwide.

During the 1849 California Gold Rush, an entrepreneurial group of travelers from Georgia was taking a band of horses west where they were needed. The group hit the Rocky Mountains too late in the season to make a feasible crossing and was forced to winter near present-day Denver. As they were veterans of the Auraria, Georgia, gold rush of the 1820s, they knew how to prospect for gold and found some good quantities in streams near where they were camped. At this time, Indian dangers were real and prevented them from going into the mountains to explore further. Springtime found them once again en route to California, where they sold the horses and undertook their own mining adventure in the goldfields. They all agreed to return to the Rocky Mountains to search for gold there at a later date.

In May 1858, the group of seven met in St. Louis, reaching Denver by August and began prospecting. W. Green Russell discovered rich placer deposits at Russell Gulch while others of the party returned to the East to get more supplies. In 1857, the war department sent an expedition in the West to find the Cheyenne Indians who had been on the warpath for a year. The soldiers were to bring the Indians to terms or to chastise them. One of the parties made its way to Cherry Creek near Denver and, while marching along, ran into a party of Missourians. These Missourians had also discovered gold in the mountains. In January 1859, George A. Jackson had discovered placer gold along Chicago Creek near the hot springs in Idaho Springs and was very secretive about his discovery. He noted the location and returned in the spring. By then, he had been followed, and the rush was on. Meanwhile, in May 1859, John Gregory of Georgia had worked his way up North Clear Creek and discovered the source of lode gold in Black Hawk. That summer brought thousands of new prospectors to Clear Creek and Gilpin Counties.

"Pikes Peak or bust!" was the motto used by over 100,000 fortune-seekers undertaking the long journey to Colorado from "the States." (Pikes Peak is located nowhere near the Kansas gold discoveries. Zebulon Pike's expedition sighted Pikes Peak in 1806. The area was remote territory and often referred to as "Pike's Peak country.") Word spread quickly; the New Eldorado and "Gold in Kansas Territory" were the cries as a desperate nation sought relief from the financial panic of 1857.

As most of the good claims were taken at Central City in the summer of 1859, George and David Griffith, two brothers from Kentucky, continued to prospect farther up the creek until they discovered a small amount of gold at the future Georgetown. They established a small community there with their families. This area eventually became the Griffith mining district and was not fully developed until 1864, when men from the Union district discovered the Belmont Lode on Mount McClellan, a massive silver deposit located above Georgetown.

In 1860, prospectors from Nevadaville, Gilpin County, continued seeking more mineral wealth over the mountains until they came upon gold discoveries at what would become North Empire (Upper Empire). This led to placer activity farther below, along Clear Creek at Empire. Empire became a supply point for Upper Empire and eventually became the Upper Union mining district. Many left Idaho Springs for the promise of the new discoveries at Empire. They also fanned out and came over the mountains to form mining districts above Fall River and in every valley between Payne's Bar (Idaho Springs) and Georgetown. Eventually, 19 mining districts were set up. Each district had its own set of laws to protect the miners from trespassers and thieves and its own set of rules for mining.

The pioneer settlers established communities up and down the Clear Creek Valley. The mines of Clear Creek County provided a way of life for the pioneers. The pioneers brought the East to the West. They felt as if they were setting up a new way of life, much like the pioneers of the American Revolution. Life was hard, but they were able to build up communities that they could be proud of. The cycle of gold and silver did not last, but it left a mark on Clear Creek County for generations to come. The communities were established to serve the mines, and in turn, the mines served the communities for many years. These years of productivity remain evident in Clear Creek County and have permanently defined the local communities.

Note to readers: It is very dangerous to enter old, abandoned mine tunnels, and one should not undertake the risk and danger—leave them alone. Curiosity can get you killed; please keep out and respect Private Property signs.

One

Pikes Peak or Bust!

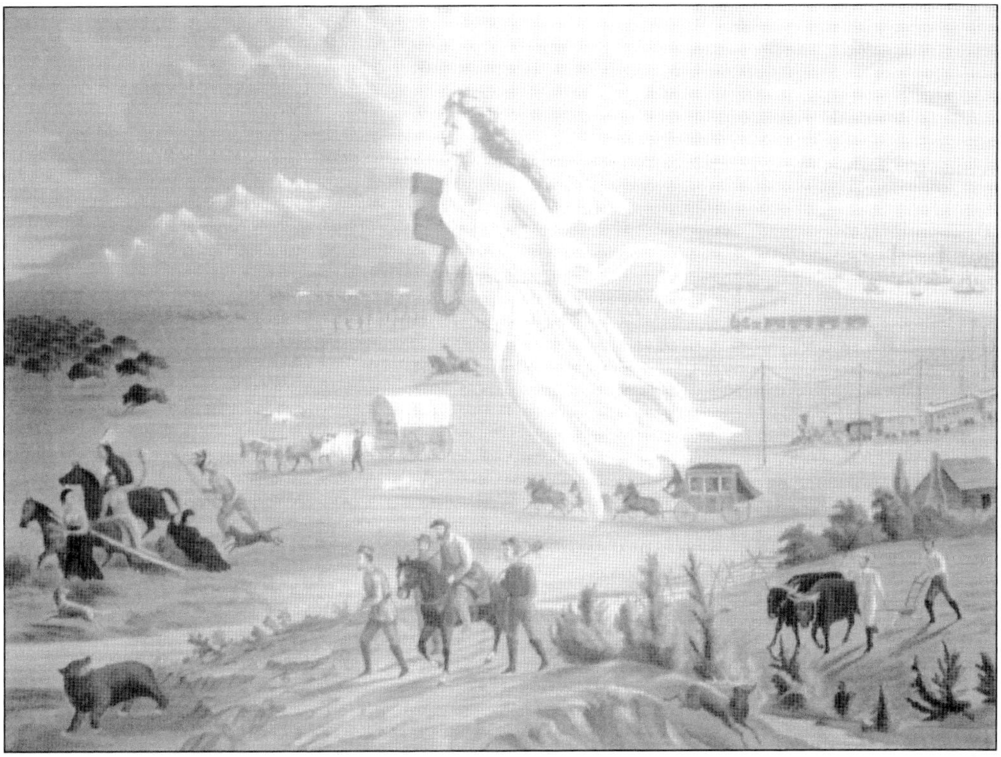

The 1873 painting is titled "*American Progress*. As America grew during the 19th century, the concept of Manifest Destiny was prevalent. It was the belief that the American civilization should expand westward and dominate the continent from coast to coast. In the painting, Columbia (representing the United States) strings telegraph wire in her journey west. John L. O'Sullivan wrote on the concept of Manifest Destiny in 1839 with the article "The Great Nation of Futurity" as part of the Making of America series at Cornell University. This gave cause to Pres. James Polk, who, at his inaugural address, stated that the government should try to extend itself over a vast territory. During his presidency, he involved the United States in the Mexican-American War. The public supported him philosophically, and events, in turn, would lead to a US expansion westward. (LOC.)

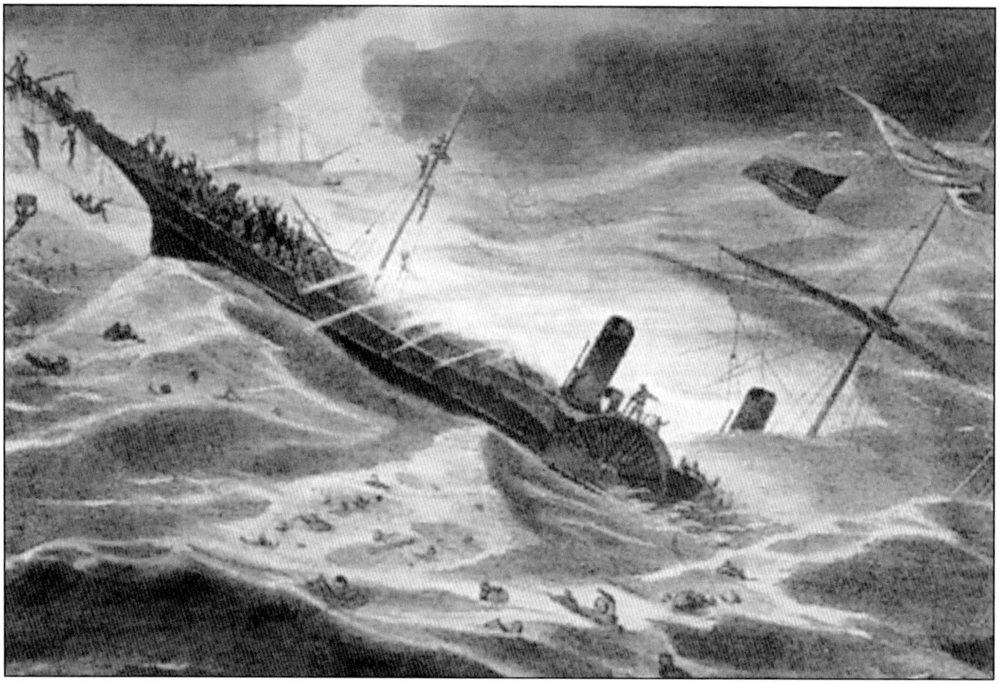

The sinking of the SS *Central America* single-handedly almost destroyed the financial well-being of the United States in September 1857 when the boat sank in a hurricane. The ship was carrying 550 passengers and 30,000 pounds in gold valued at $2 million. The loss shook the public confidence in the economy, which was a contributing factor that led to the panic of 1857. (LOC.)

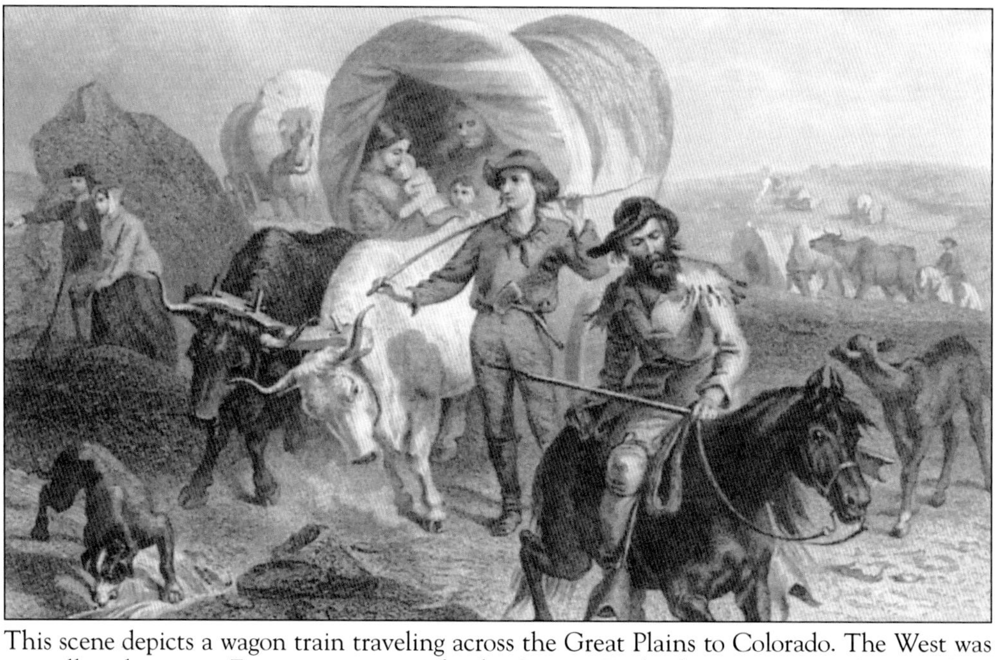

This scene depicts a wagon train traveling across the Great Plains to Colorado. The West was virtually unknown to Easterners—except for the Oregon Trail. There were many dangers along the way. Once families made it to where they were going, difficulties abounded. Many pioneers used the canvas from their wagons to build makeshift tents. (LOC.)

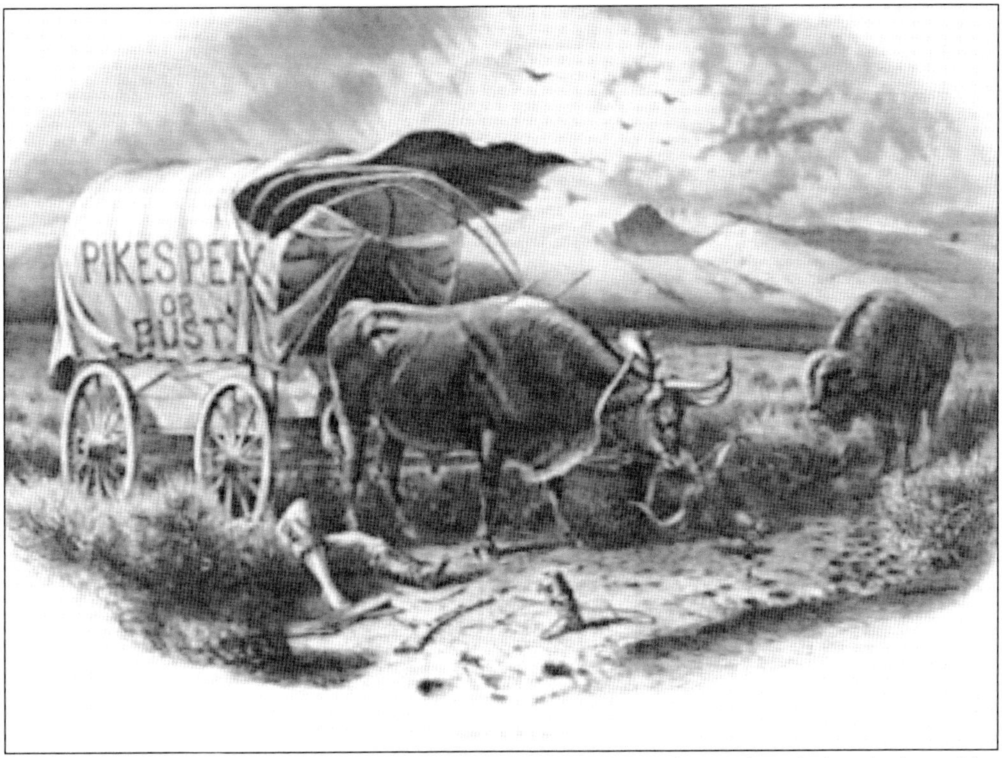

"Pikes Peak or bust!" was the motto for many travelers along the trail to Colorado from "the States." Colorado was in Kansas Territory, and many travelers endured hardships along the way, including hunger, Indians, and rough terrain. Many wound up like this and did not make it to their final destination. Times were desperate. (DPL.)

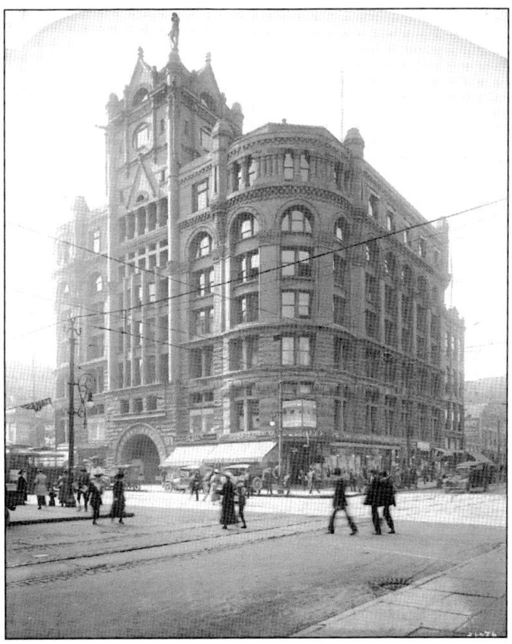

Colorado mining overcame many obstacles early on and was quite solid by 1891, when the Colorado Gold Mining Stock Exchange Building was constructed in Denver. Col. John Straughn, a wheelwright from Black Hawk, was used as the model for the 12-foot-high, 490-pound copper statue that was placed atop the fanciful structure. The statue symbolizes industry, patience, and enterprise. (DPL.)

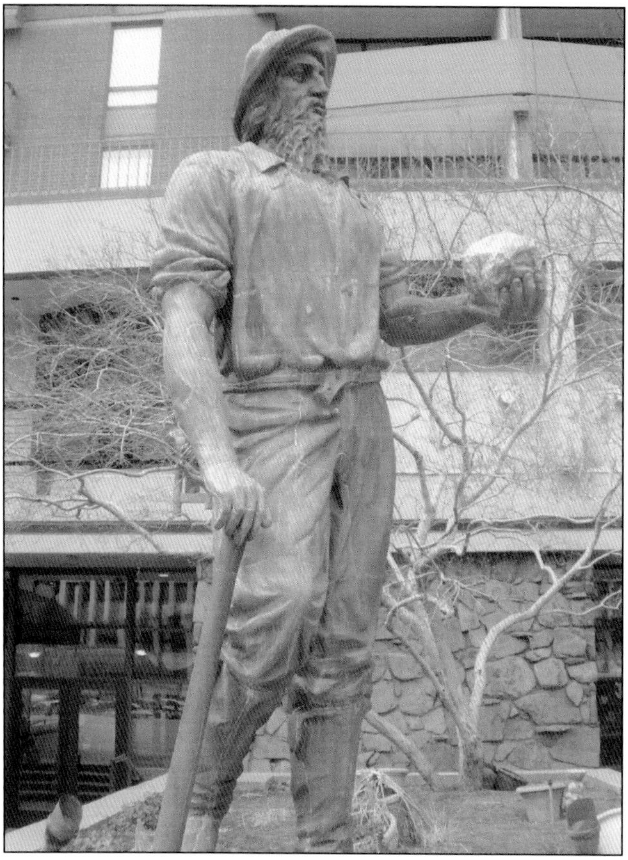

The John Straughn statue symbolizes an old prospector from Colorado's early days. The objective for any prospector was to find gold. In addition to being a wheelwright, Straughn was a prospector as well as a Civil War veteran. His wheelwright business in Black Hawk was in the present Bull Durham Casino Building. The statue was moved to its present location in front of Brooks Towers in 1961, when the mining building was torn down and Brooks Towers was constructed on the site. (Jim Prochaska.)

Colorado mining evolved from the abilities of the single miner to the need for corporate investment to develop and properly work the mines. Small operations, such as this one, did not endure. Many corporations were formed in the East for the purpose of seeking investments to run and operate the mines.

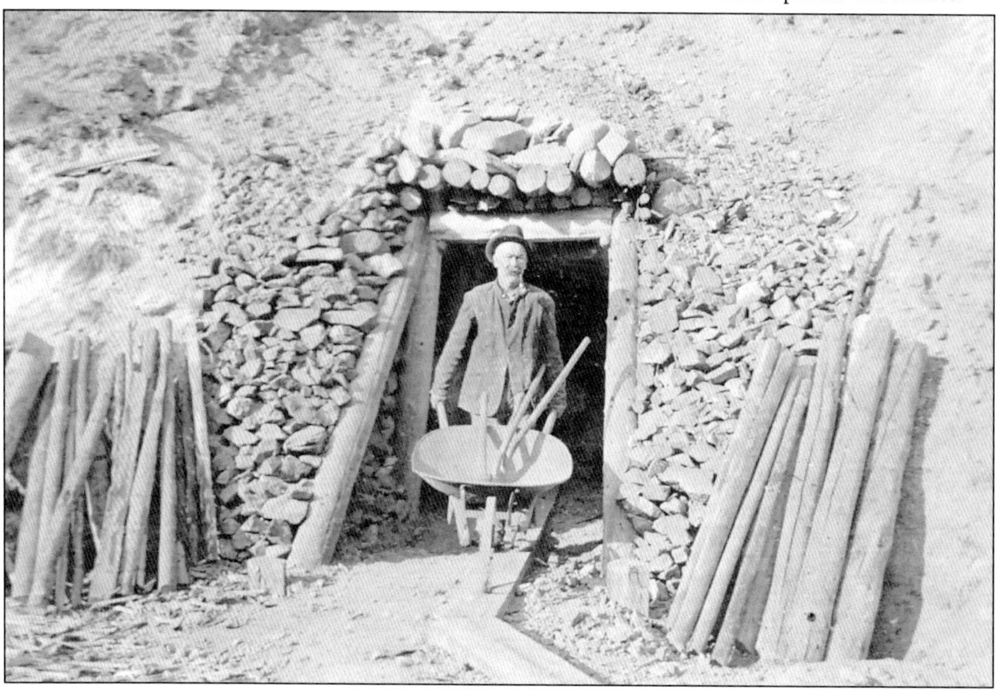

Two

Idaho Springs, Lamartine, and Alice

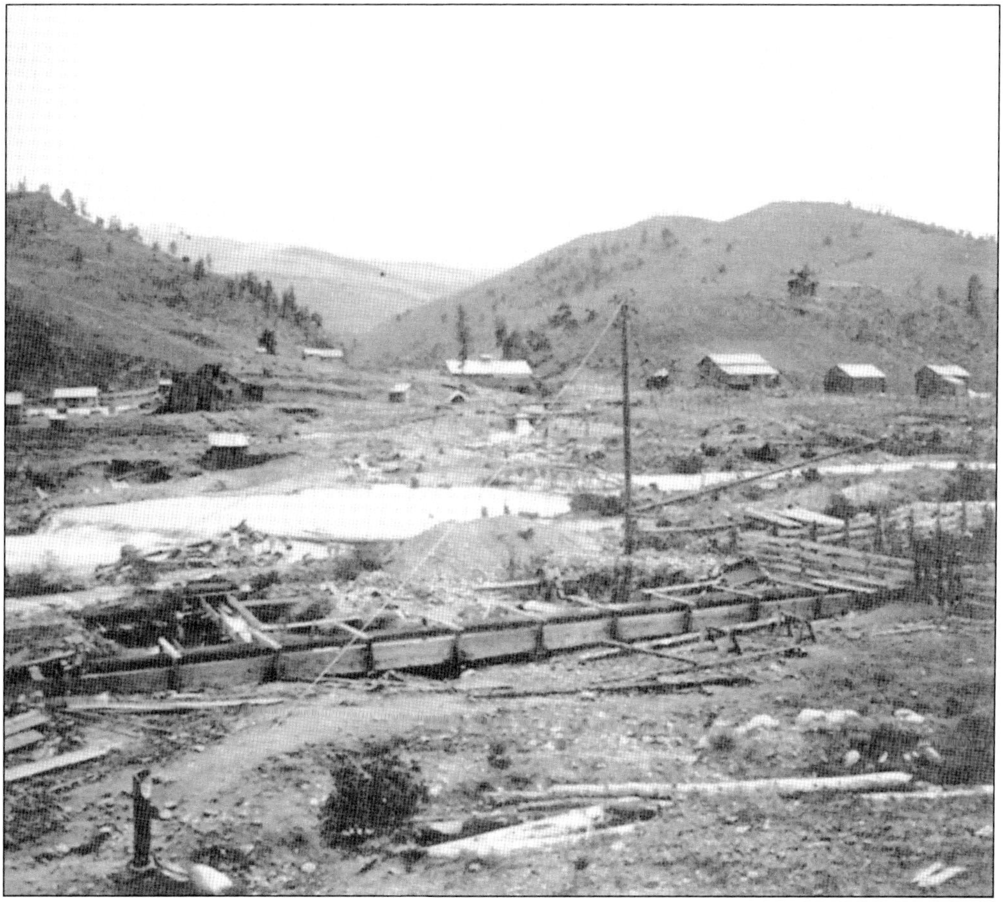

In January 1859, George Jackson first discovered gold in Idaho Springs. The area was in its early developmental stages, as this photograph depicts. The dams were enlarged, and several mills were built in this area. (DPL.)

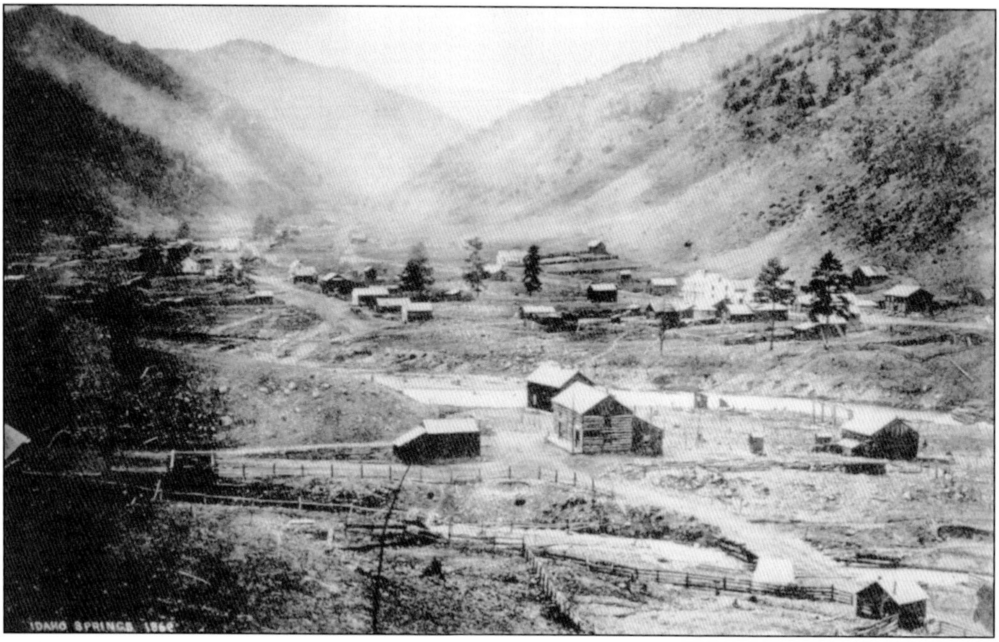

Idaho Springs is pictured here in 1869. The Beebe House can be seen in right center. This was an early pioneer hotel where Gen. Ulysses S. Grant and his entourage stopped on the way to a Georgetown visit in 1868. Stagecoach driver Bill Updike took General Grant over the Virginia Canyon road so fast that all he kept saying was "oh my God." The road is referred to as the Oh My God Road to this day. (DPL.)

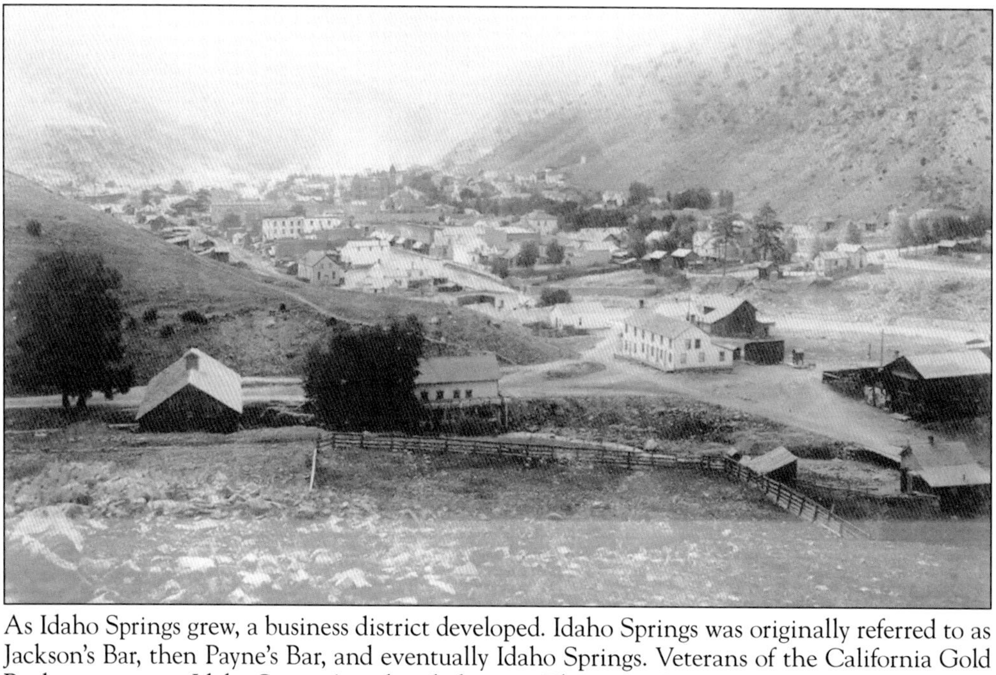
As Idaho Springs grew, a business district developed. Idaho Springs was originally referred to as Jackson's Bar, then Payne's Bar, and eventually Idaho Springs. Veterans of the California Gold Rush were among Idaho Springs's early inhabitants. The name Sacramento was used for a short time, and the Grass Valley mining district certainly has origins in veteran miners of the California Gold Rush of 1849. (DPL.)

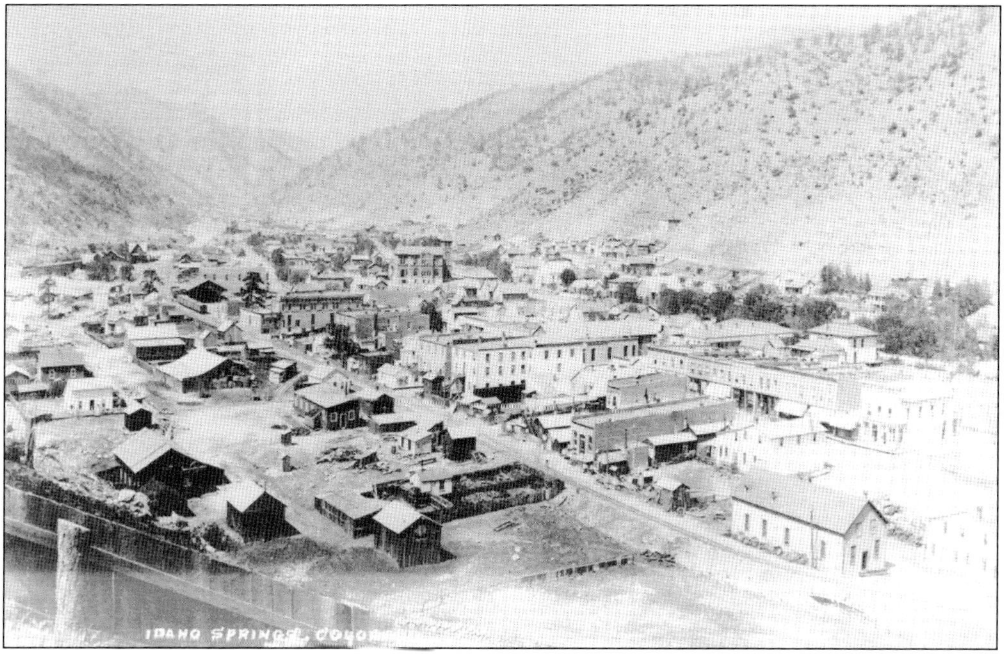

In the 1890s, as prosperity continued for Idaho Springs, more and more permanent buildings were constructed, including a new school building (upper mid-left center), the Metropolitan Hotel (mid-center); Clear Creek Foundry & Machine shop directly behind; a wagon shop (four windows along railroad tracks); and the German House (far lower right, eight windows), and the train depot (left of center). The railroad continued up the Clear Creek Valley, carrying ore and passengers along the way. (DPL.)

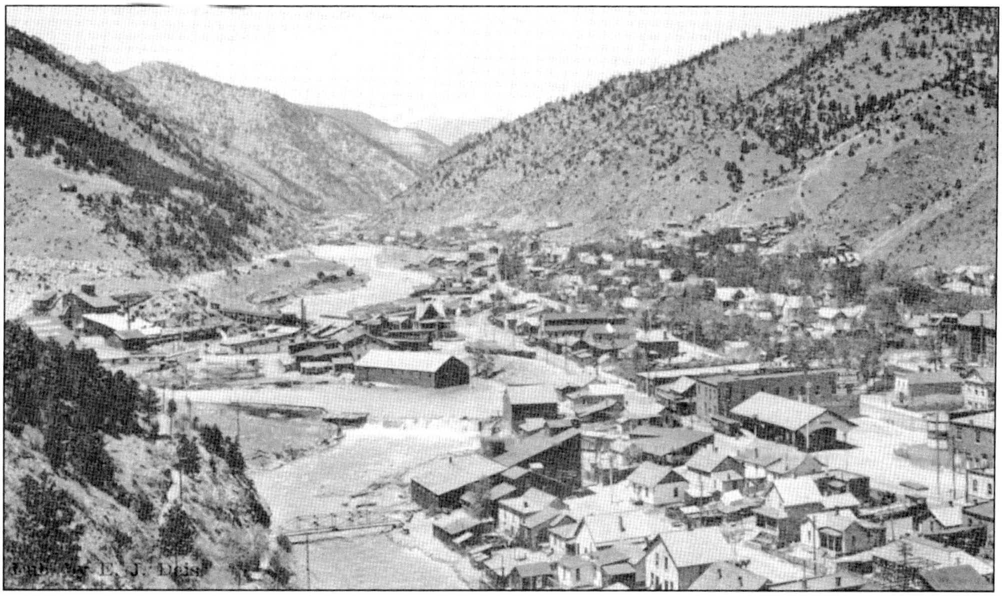

The area where Jackson discovered gold has been built up with the addition of several mills and ore reduction companies. The Jackson Concentrating Works is at far left of center. The Mixsell Mill and Dam can be seen left of center. The railroad depot is at lower right of center. The school is visible at far right of center.

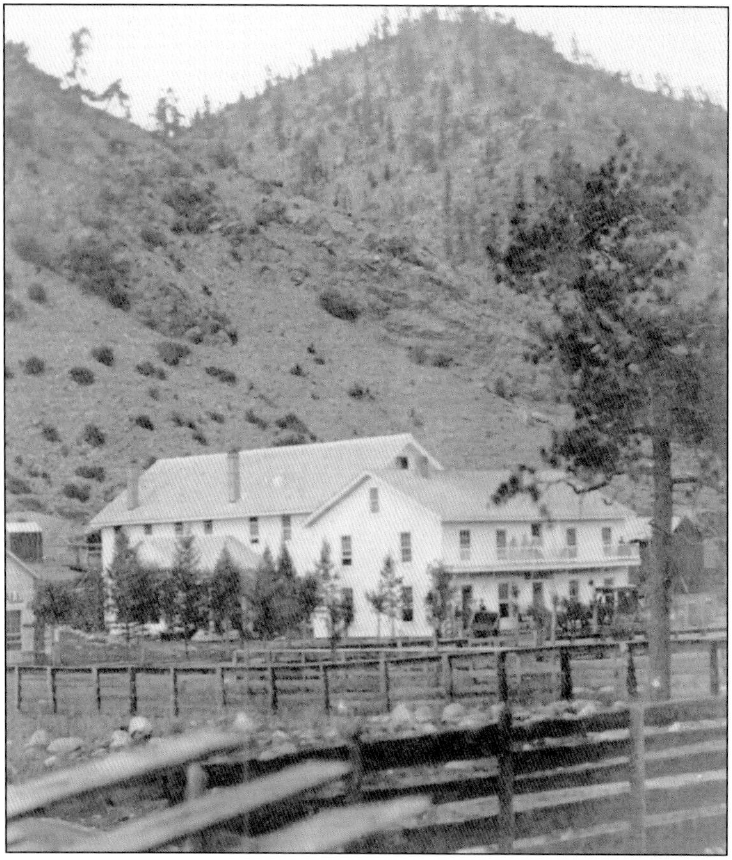

The Beebe House in Idaho Springs was once a stopping point for the party of Gen. U.S. Grant en route to the mines of Georgetown. (DPL.)

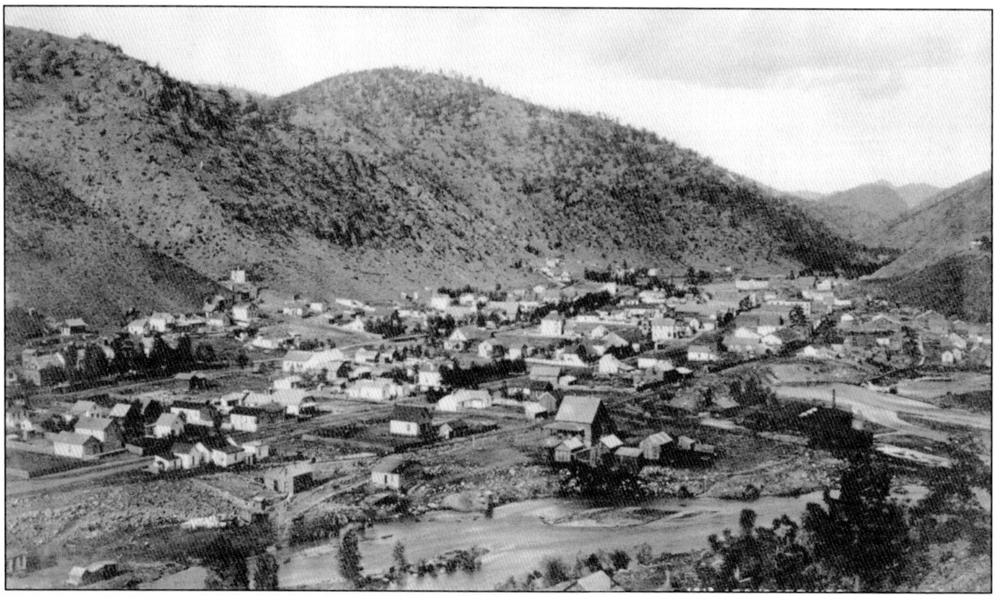

A view of Idaho Springs looking east shows Clear Creek and the Bertha Mill. The mill had direct access to the Colorado & Southern (C&S) Railway to ship its concentrates to a smelter for further reduction. (DPL.)

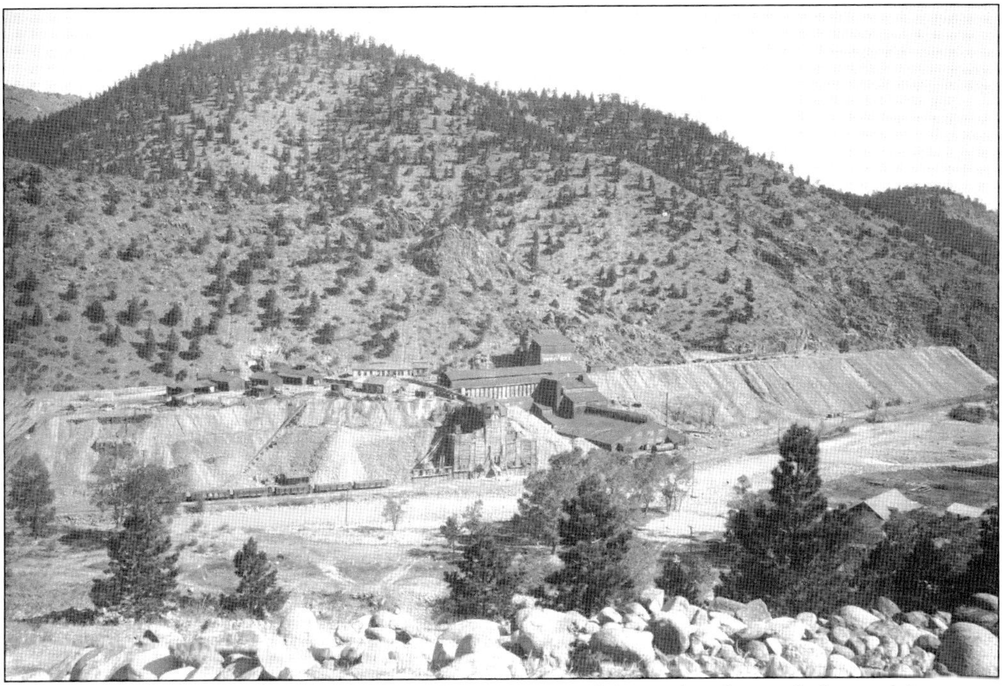

Here is a view of the Argo Mine and Mill. The Argo Tunnel, located underneath Seaton Mountain, was blasted through to Central City from 1893 to 1912 to help drain and service the mines of Gilpin County. The mill was built in 1913. Work was stopped on the tunnel until contracts were signed with the mining companies on the other side. (UCB.)

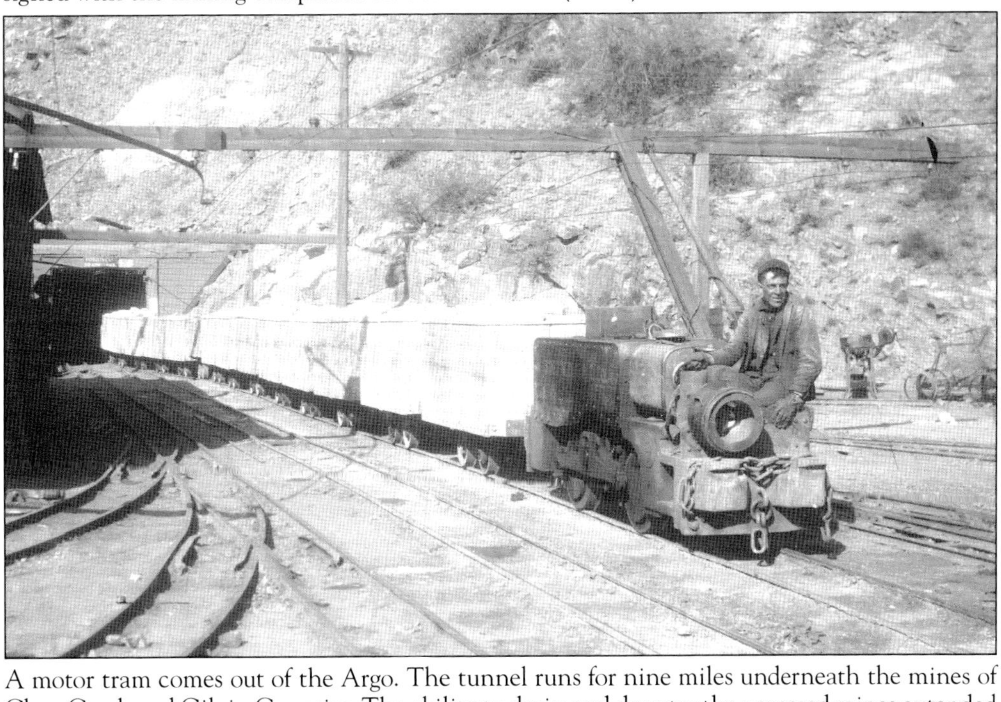

A motor tram comes out of the Argo. The tunnel runs for nine miles underneath the mines of Clear Creek and Gilpin Counties. The ability to drain and dewater the accessed mines extended their useful lives by several years. (UCB.)

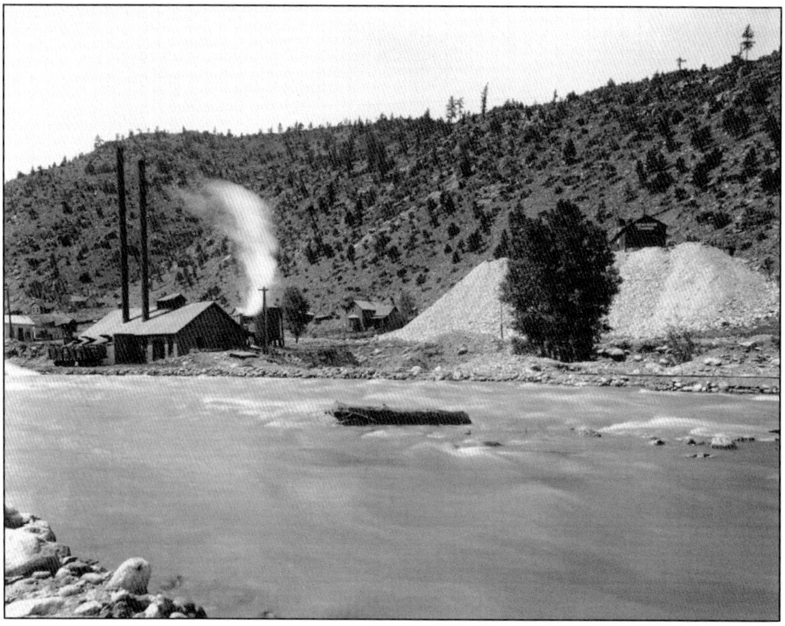

This image captures the early beginnings of the Newhouse Tunnel around 1900. The Sun and Moon powerhouse (left center) sits along Clear Creek. Note the ore car for the C&S Railway line that runs along Clear Creek. Compare this photograph to the one on the opposite page. (UCB.)

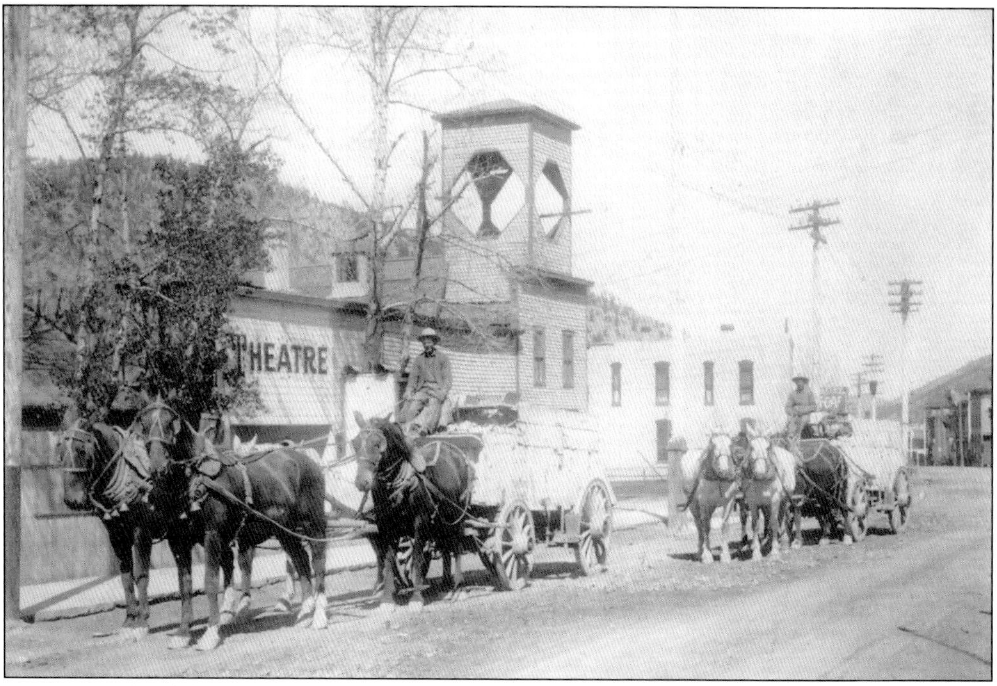

Wagons are loaded with ore from the Two Brothers Tunnel. (See photograph at bottom of page 20 for the Two Brothers operation.) These wagons were very heavy when coming down the mountainside. It took a professional driver to get the job done. Note the theater previously used as an opera house. In 1877, to coincide with the arrival of the railroad, the new state of Colorado asked Idaho Springs to put on drilling and mining contests open to the entire state for the local Fourth of July celebration that summer. This would showcase the newly arrived railroad. A stage remained and was built over and converted into an opera house. It was utilized for a few years until a more permanent building was constructed of brick on Miner Street. (UCB.)

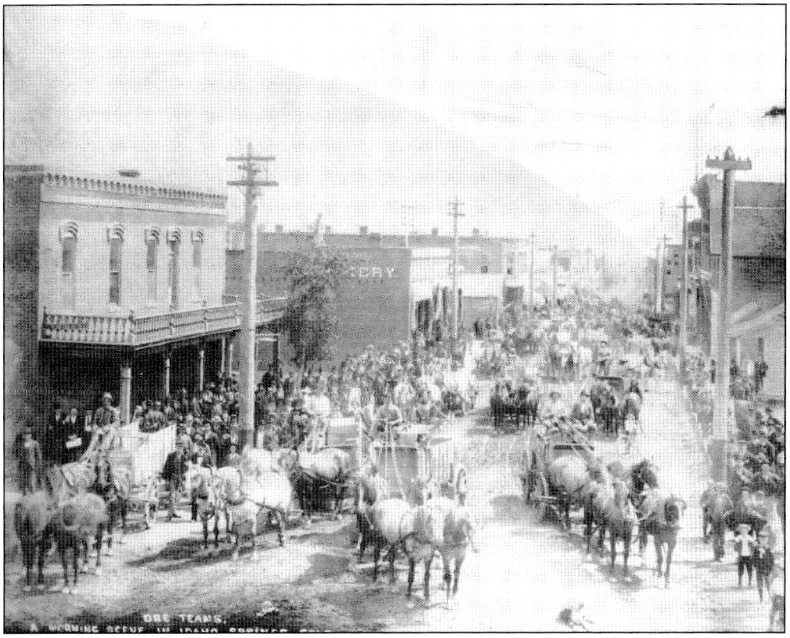

Ore wagons from the different mining operations crowd the streets in this picture. Each mine would have its own wagon team and would also compete for unloading space at the local ore sampling works. (CCHS.)

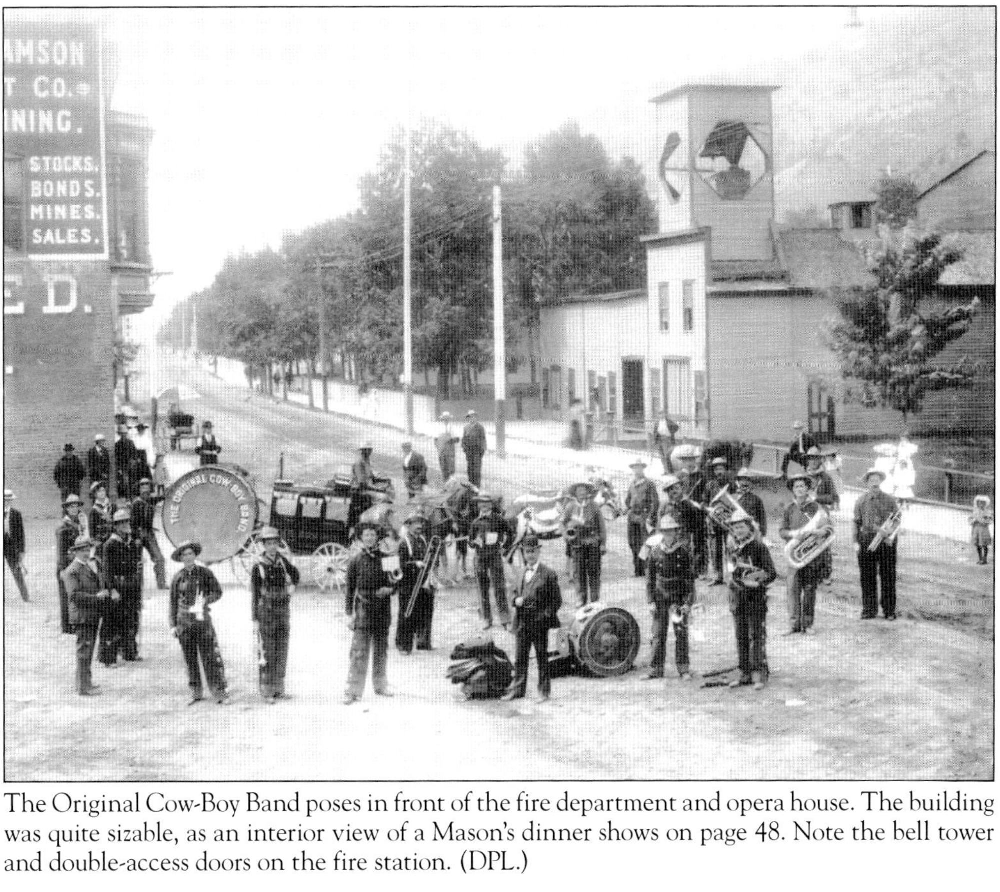

The Original Cow-Boy Band poses in front of the fire department and opera house. The building was quite sizable, as an interior view of a Mason's dinner shows on page 48. Note the bell tower and double-access doors on the fire station. (DPL.)

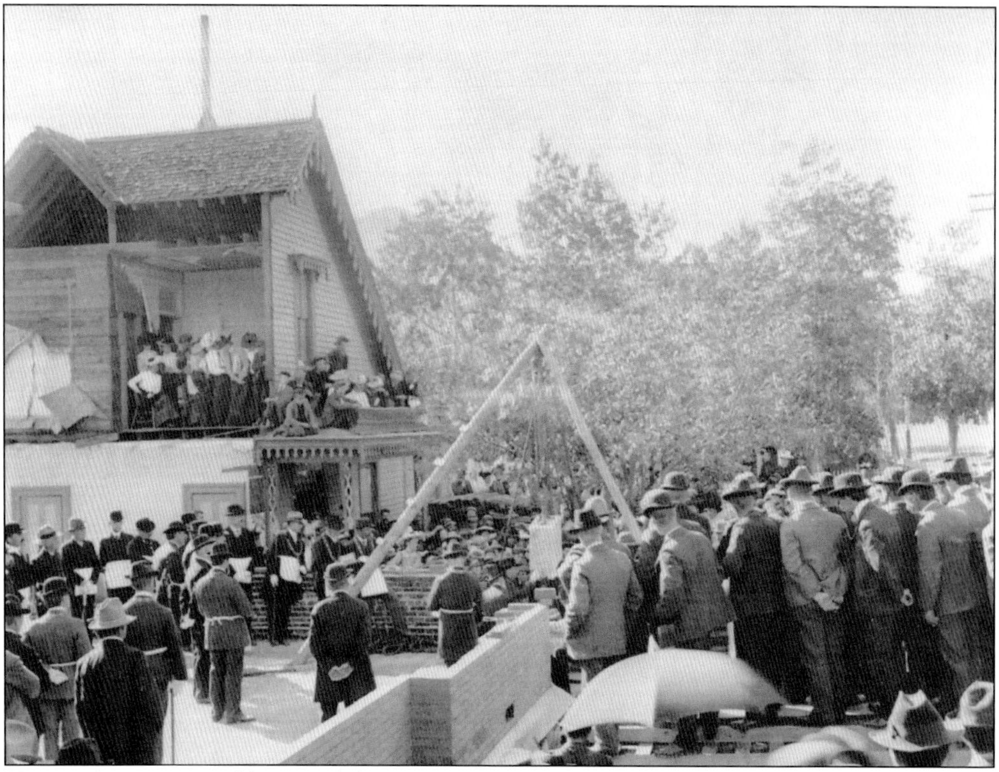

Here is the ceremonial laying of the cornerstone for an addition to the new school building in 1886. The house being torn down is the old Doe house. The parents of Baby Doe's first husband, Harvey, lived there. The Does owned the Fourth of July Mining in Central City. Baby Doe and husband, Harvey, lived in a room under the railroad in Black Hawk before she left him for H.A.W. Tabor, who made a fortune in the mines at Leadville. (DPL.)

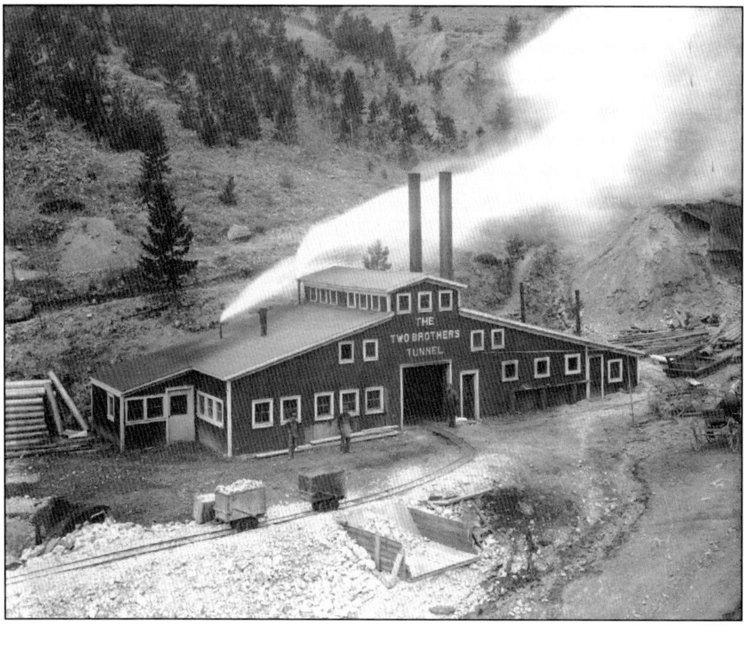

Pictured here is Two Brothers Tunnel in Virginia Canyon. Note the ore chute used to unload the ore into wagons for travel down to the mills and sampling works in Idaho Springs. As the tunnel grew, an improved ore chute system was devised. (UCB.)

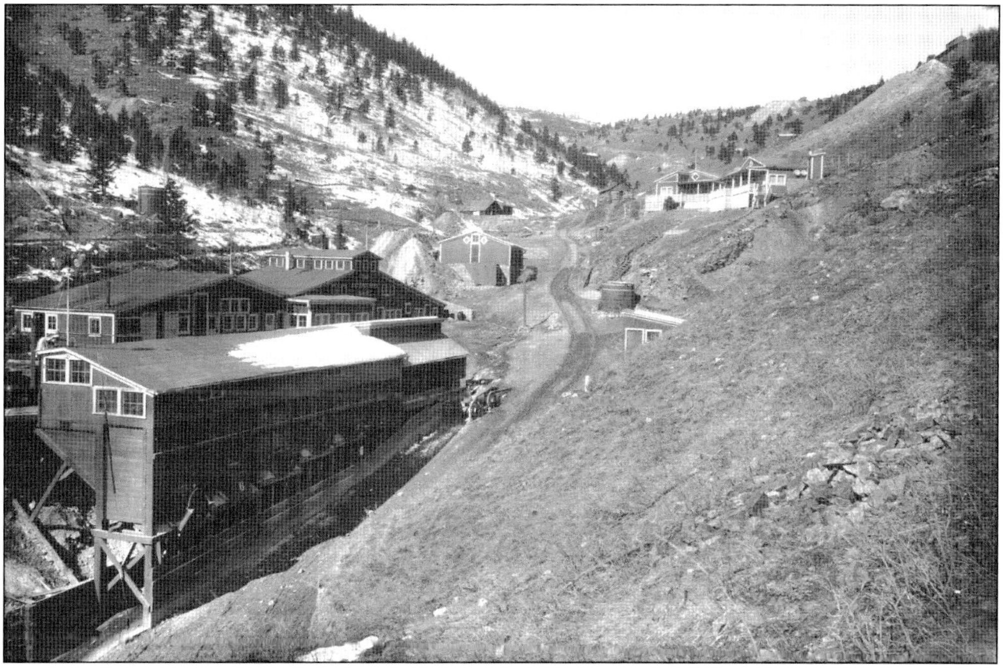

The Consolidated Two Brothers Mining operation was able to raise more funds because of the tunneling law of 1872. Tunneling was a lucrative business and usually paid dividends to all investors. Note the expansion of the operation from the previous photograph. There are seven new ore chutes and an additional cyanide building. The waste rock pile to the south (left) grew bigger and bigger as more ore was pulled from the mountain. The road continues on and eventually comes out at Central City, passing through Russell Gulch in Gilpin County. (UCB.)

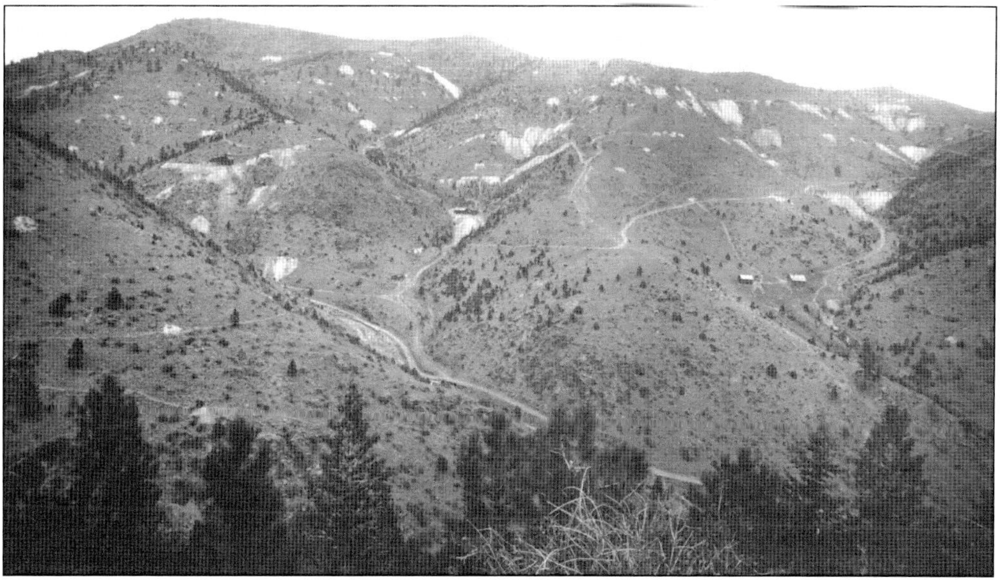

Here is a view of the mines of Virginia Canyon from across the valley. Note the road traveling through the lower center. Gilpin County is on the other side of the mountain crest. Russell Gulch was where William Green Russell discovered rich gold deposits above Central City and Black Hawk. (UCB.)

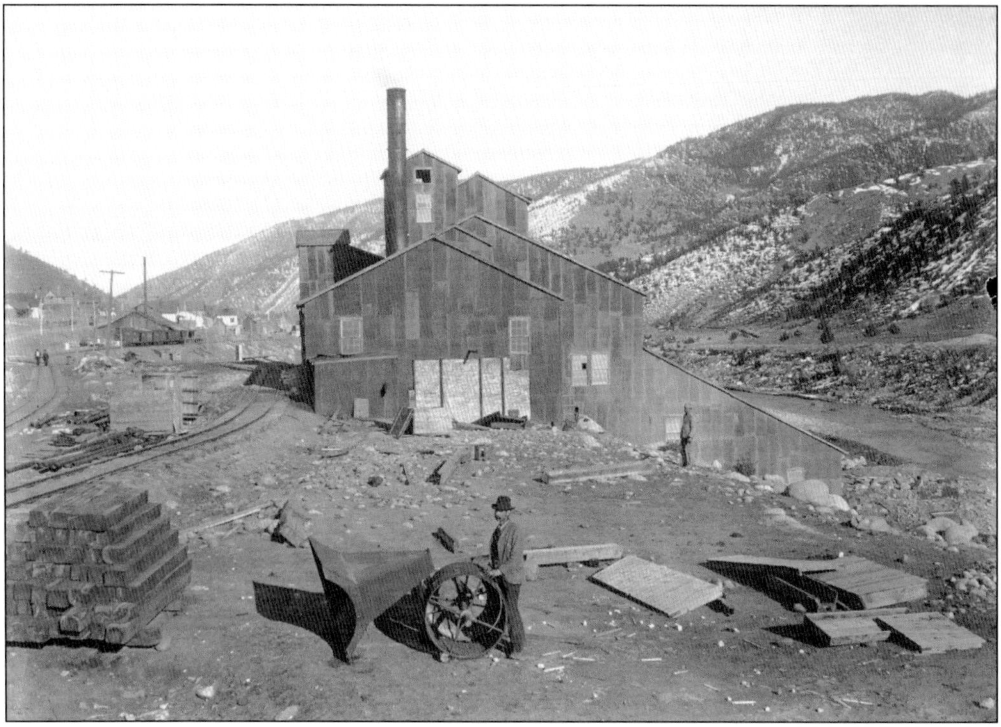

This close-up view shows Bertha Mill in approximately 1910. Railcars would be filled with processed ore to ship to the smelters. Ore from the Big Five Mine and other mines in West Idaho Springs would be processed at this location. The Colorado & Southern Railway had direct access from this mill. (UCB.)

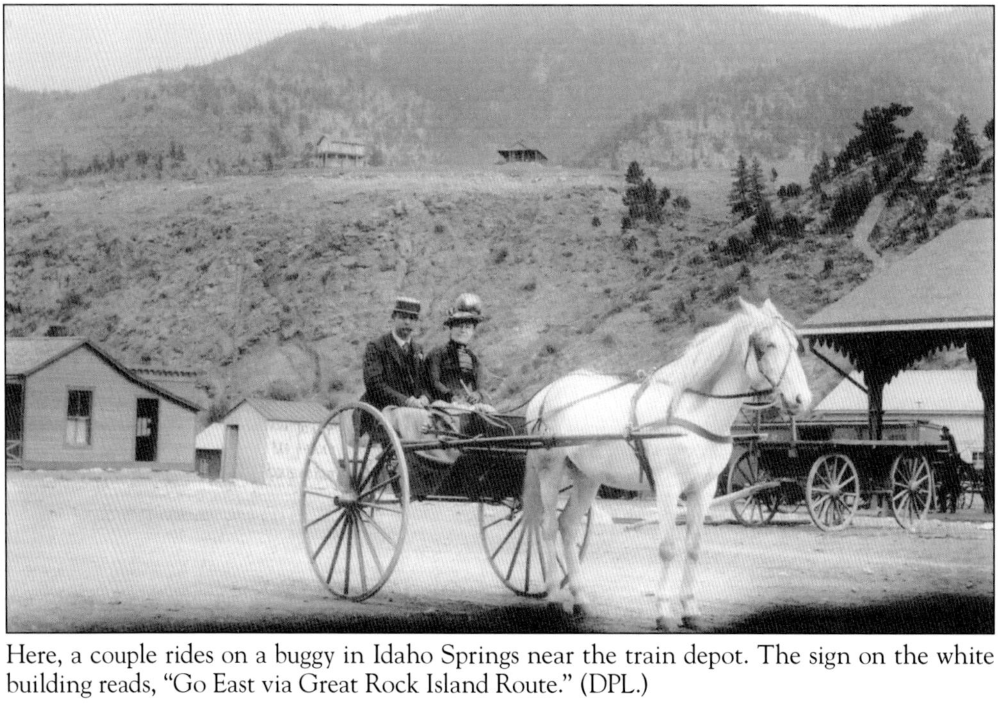

Here, a couple rides on a buggy in Idaho Springs near the train depot. The sign on the white building reads, "Go East via Great Rock Island Route." (DPL.)

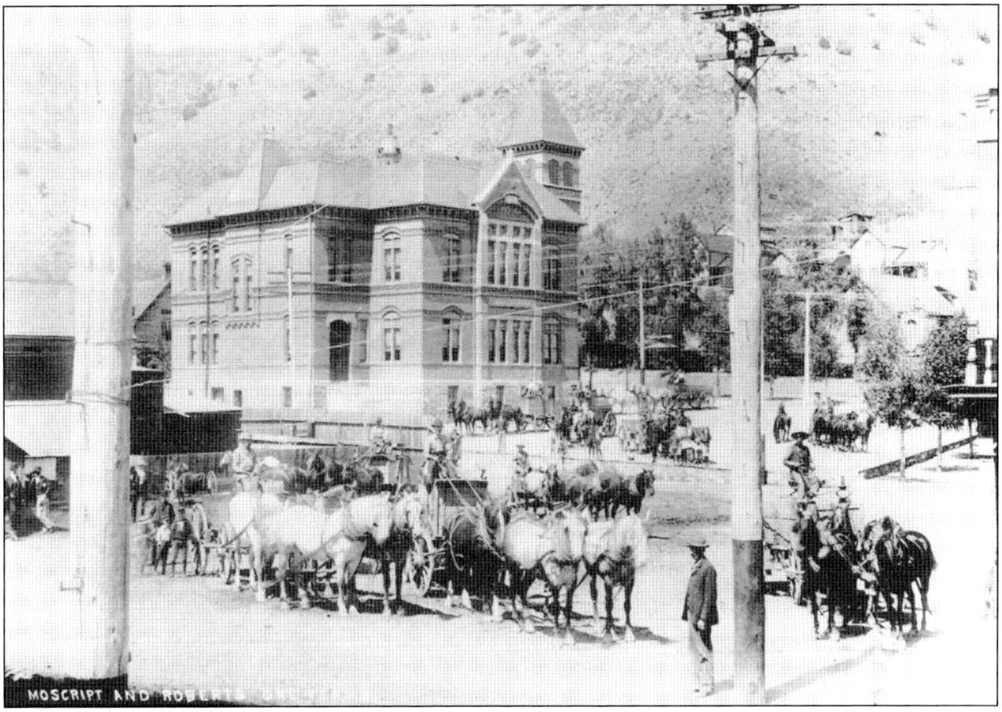

Pictured here are the Moscript and Roberts Ore teams. When a mine did not have ore wagons, it would contract with a freighting company to haul its ores to the mills. Note the Harvey Doe house left of center. This was later torn down to build an addition to the existing school. (See page 20.) (DPL.)

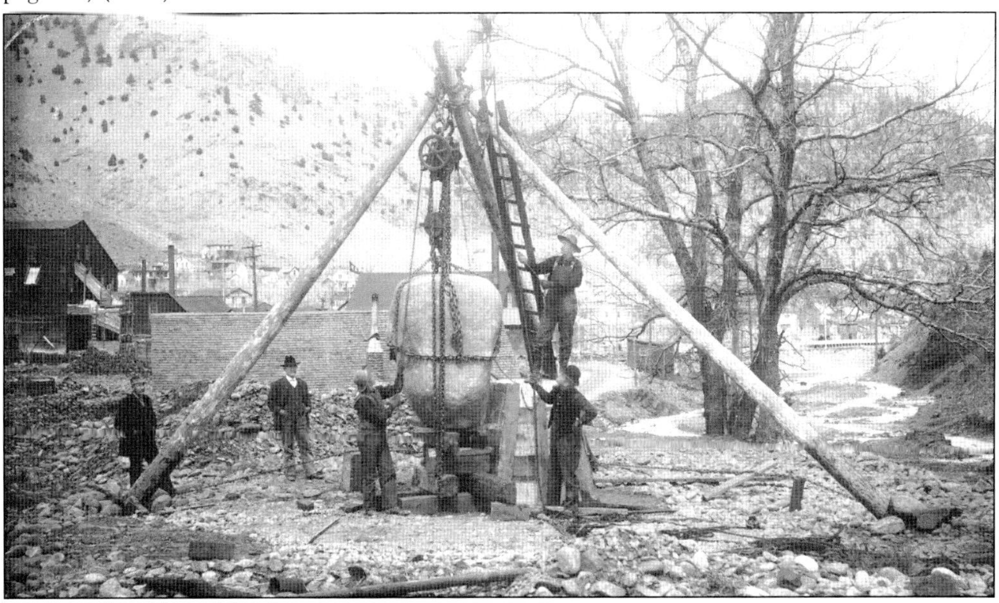
These workers are laying the foundation to the Jackson Gold Discovery marker in October 1909. Sen. Henry Teller from Central City was at the dedication. While the actual location of Jackson's discovery is on the other side of the creek, this side was chosen so as not to impede future development. (UCB.)

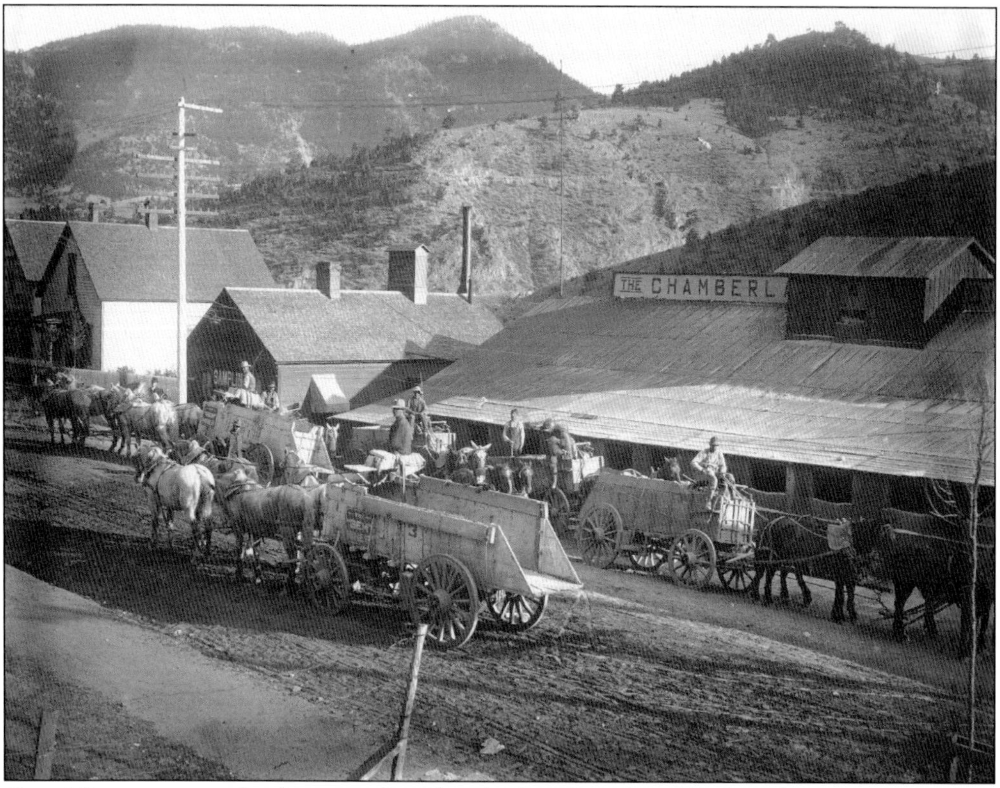

Gem Mine wagons unload ores at the Chamberlain Sampling Works. The wagons would be backed directly up to the chutes and shoveled out by hand and would then travel back up to the mine for another load. A sampling works differed from a mill. The sampling works could tell the strengths of the ores right away, whereas a mill would use different methods and take longer to discover the concentration of metals permeating in the pay rock. A mill might mix a mine's strong ores with another mine's weak ores to lessen the value of the sample. Certain measures would be put in place to prevent such deception whenever possible. An honest mill would earn a solid reputation and last for many years. (UCB.)

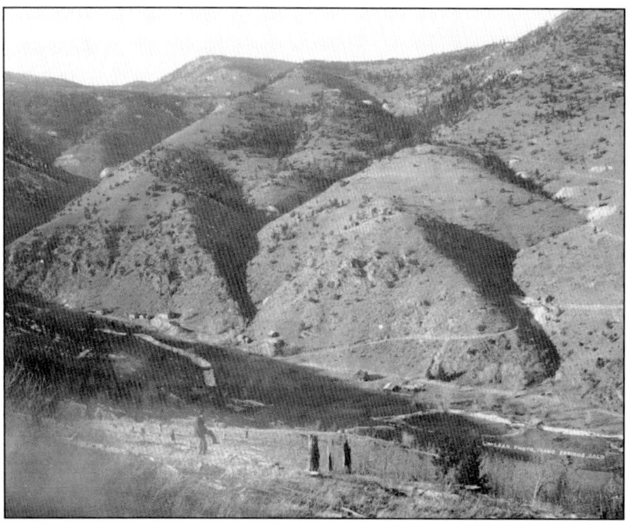

This photograph is of the Little Mattie group of mines up near Chicago Creek. Mines were named after a variety of subjects, including wives, girlfriends, small children, presidents, historical figures, famous generals, and political figures. (UCB.)

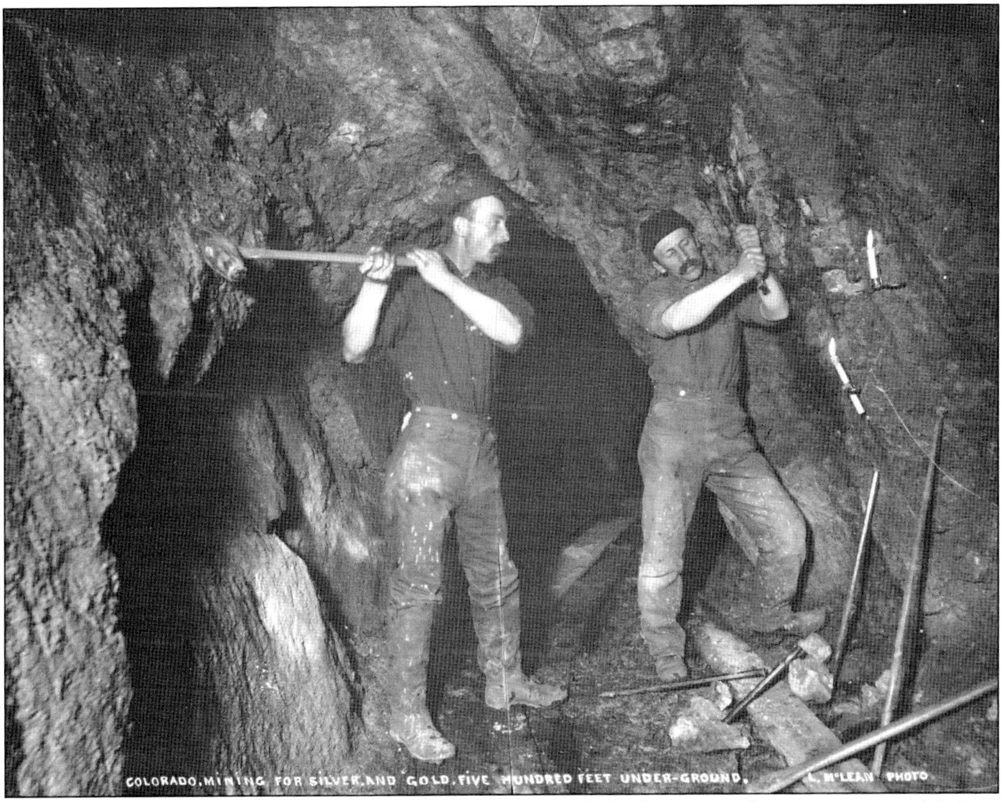

These two are double-jacking underground. In order to drill a hole into the rock, one would sling the sledgehammer and the other would turn the drill bit a quarter turn between swings. This was very dangerous, and one could easily get hurt. Partners had to take their jobs seriously and trust one another. Automated air drills came later as industrial development improved mining and milling methods. (UCB.)

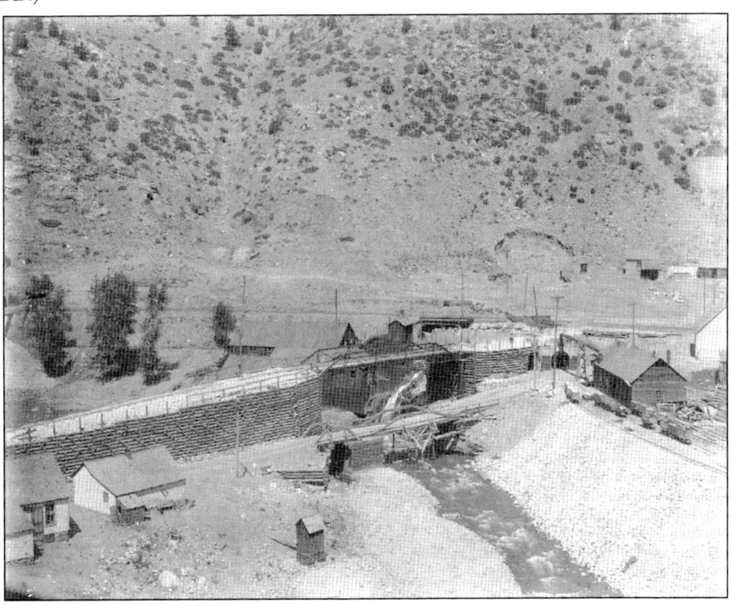

The Big Five Tunnel was located at the west end of Idaho Springs. Tunneling became a lucrative business by the 1890s. The Big Five Company owned many mines and was able to access the Lucania and the Edgar among others from this tunnel. Ores were shipped to the close-by Bertha Mill, built on the west end of Idaho Springs after the turn of the century. (UCB.)

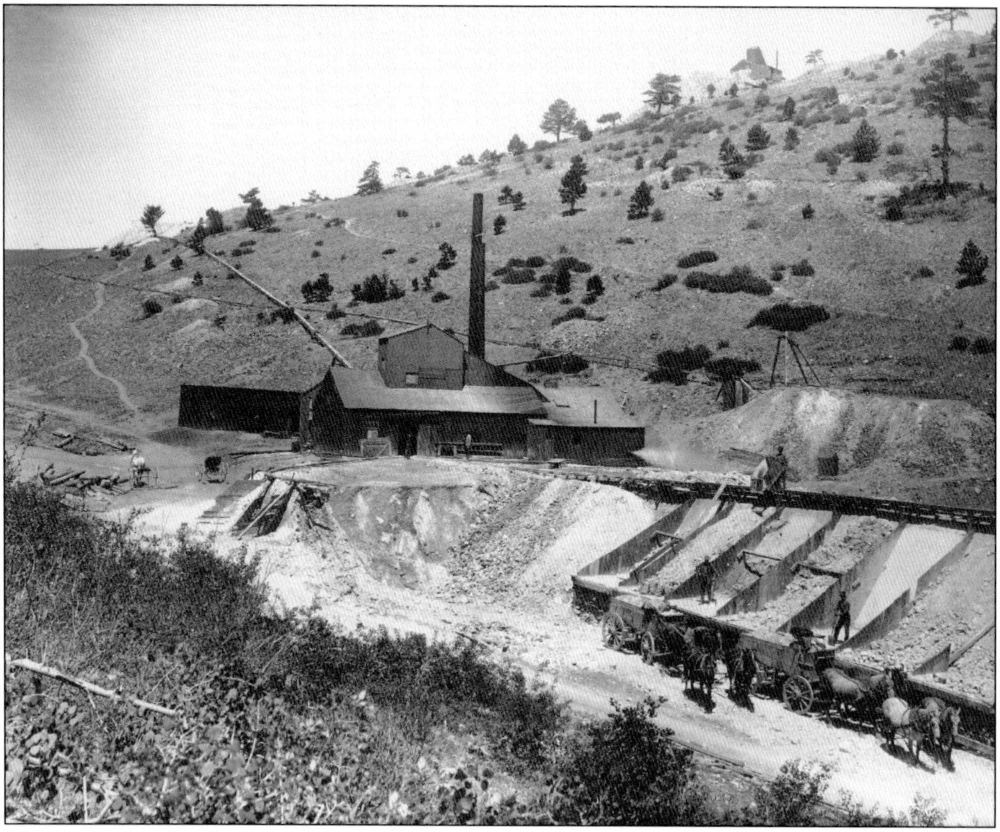

The loading and unloading of wagons at the Gem Mine was all done by hand. The powerhouse provided power for the drills and lights in the mines, replacing candles and lanterns.

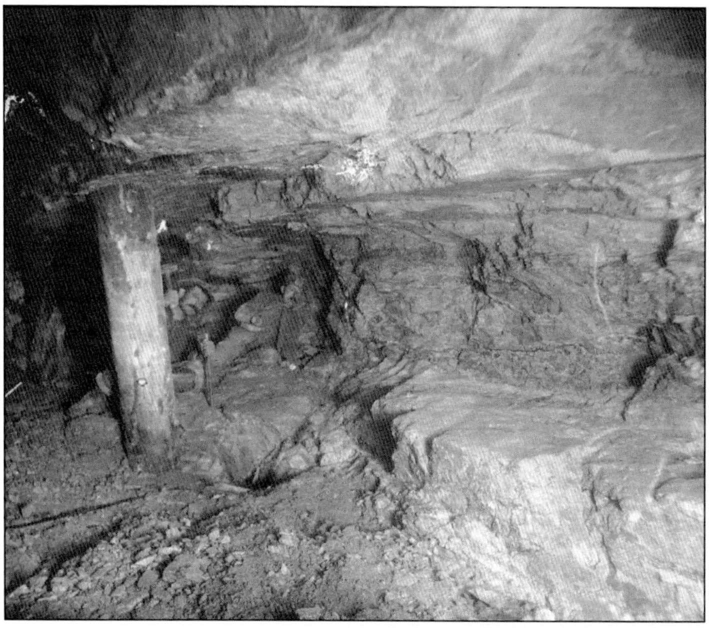

Here is an underground view of single-jacking a stope (a vertical shaft in a mine tunnel). Conditions underground were very dark and dangerous. Countless miners were blown up from a missed shot of dynamite in mines all over the county and state. (UCB.)

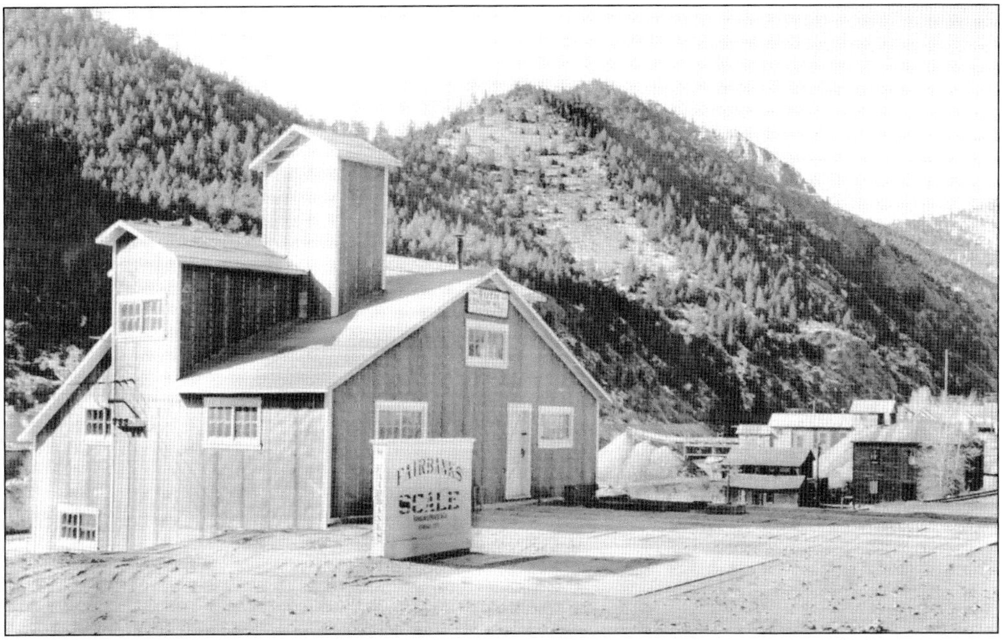

The Ruth Mill at the west end of Idaho Springs had a scale upon which to weigh the incoming ore. The Fairbanks Scale was invented by the Fairbanks brothers in the 1830s and developed by the Monarch Company in Kansas City in later years. The company still exists and now specializes in truck stop scales. Following the 1980 fire at the old Kilton Mill, this site was turned into a playground. (UCB.)

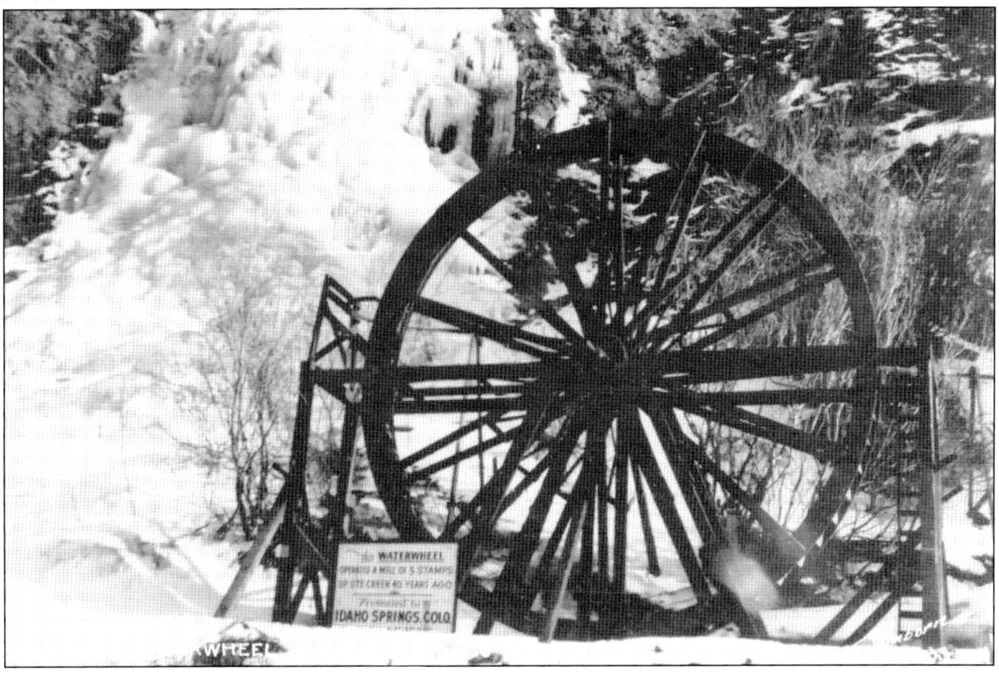

Charlie Tayler used this waterwheel at Idaho Springs to power up the Chicago Creek mine. Public service employees moved it in place in 1945. It has become a symbol for Idaho Springs; the city had the waterwheel renovated and repaired in the mid-1980s.

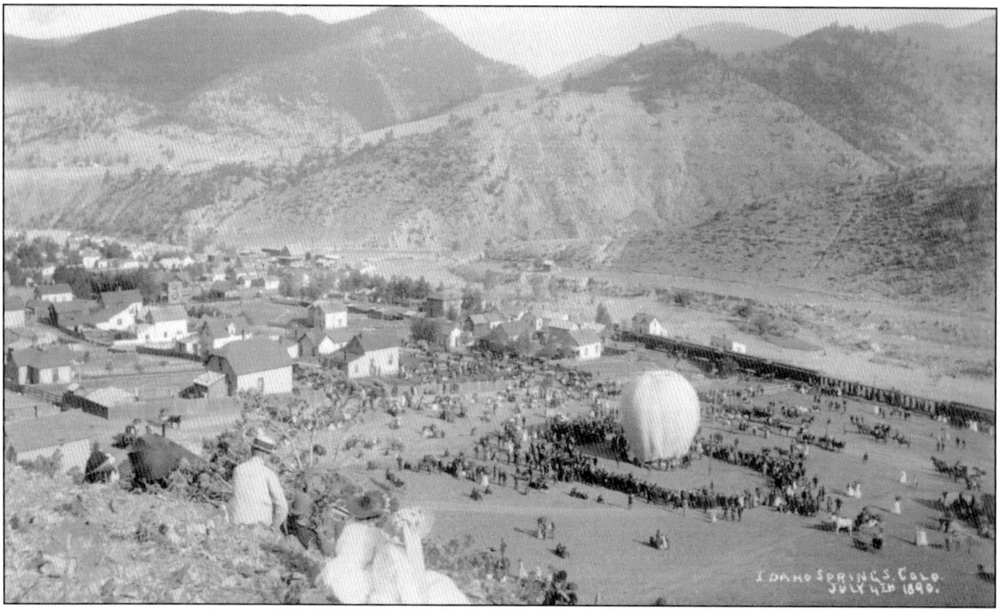

Attendance was high for raising a balloon at the 1890 Fourth of July celebration at Idaho Springs. Fourth of July celebrations in Clear Creek County always pitted the local fire crews against each other for hose cart races, rotating whose turn it was to visit each other (see page 51). (History Colorado.)

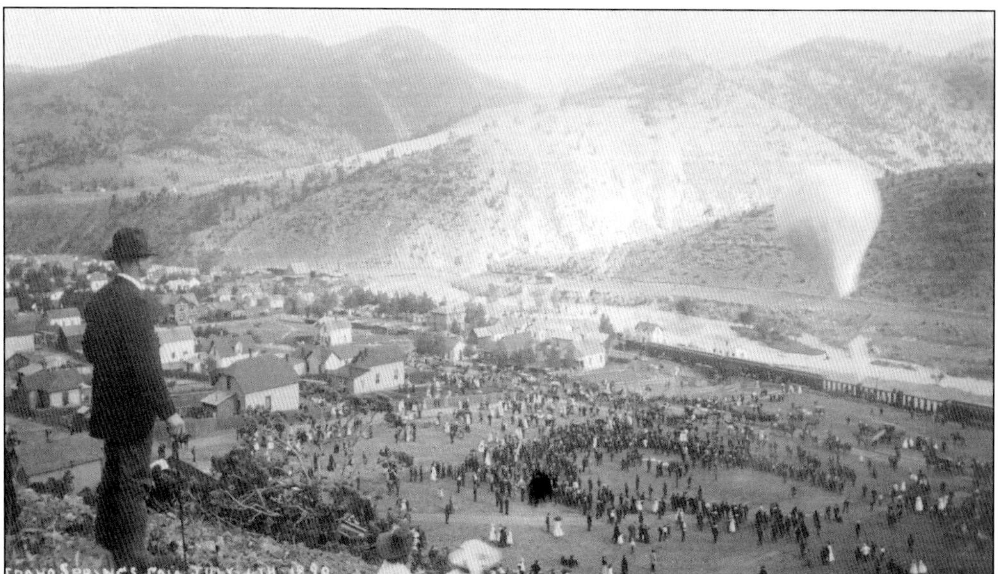

The balloon is raised. The 1890s were very popular for ballooning. Similar events were held at Georgetown and Central City. (History Colorado.)

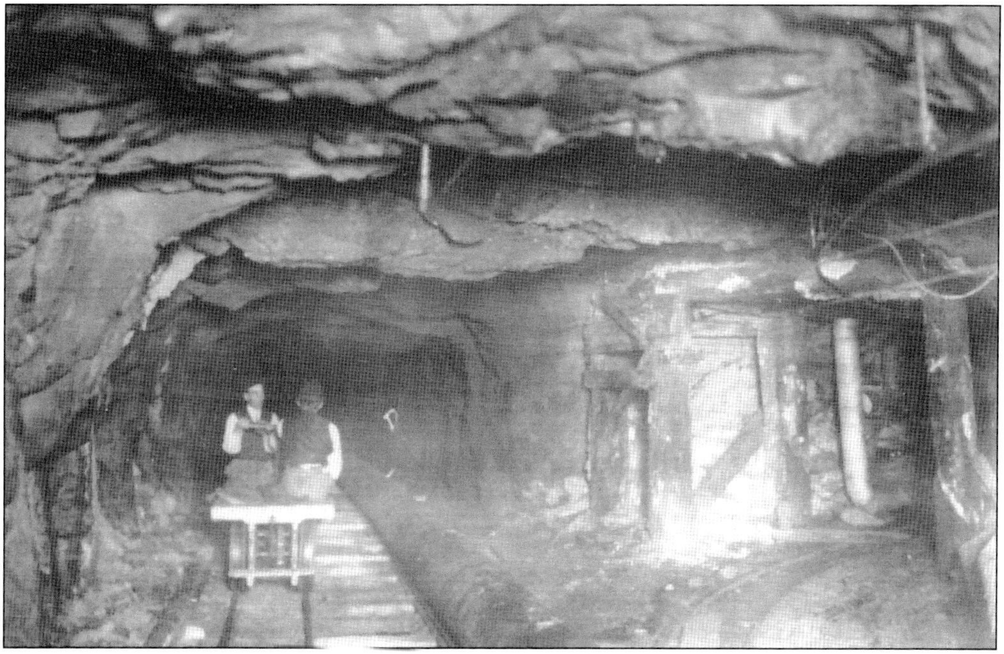

Here is an interior view of the Sun and Moon Tunnel. Labor strife in 1904 caused striking miners to blow up the mine and two other miners. They were put on trial and found guilty. Frank James said, in his time, speaking in terms of the Civil War, "If there is ever another war in this country, it will be capital and labor, I mean between greed and manhood, and I'm as ready to march now in defense of American manhood as I was when a boy in defense of the South." Nationwide violence and upheaval caused drastic changes in the rights of miners between 1894 and the 1920s. In Ludlow, Colorado, in 1914, approximately 20 miners were killed in a raid on the tent colony of the striking Colorado Fuel & Iron (CF&I) workers. Matewan, West Virginia, was the site of another violent encounter between labor and management in 1920. (DPL.)

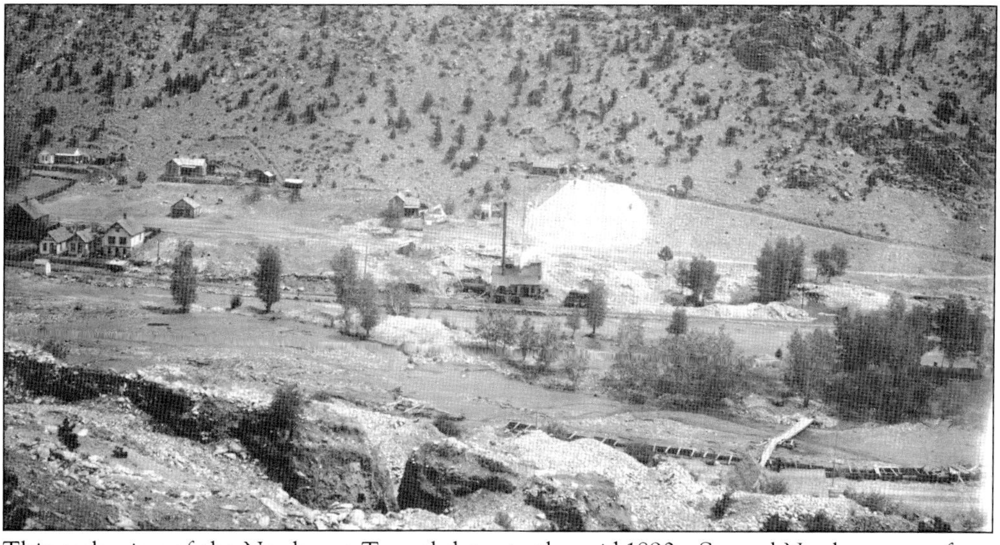

This early view of the Newhouse Tunnel dates to the mid-1890s. Samuel Newhouse was from Philadelphia and became wealthy from the Leadville mines. He organized the Newhouse Tunnel with investment money from Chicago and Boston, and the tunnel made him very rich. (UCB.)

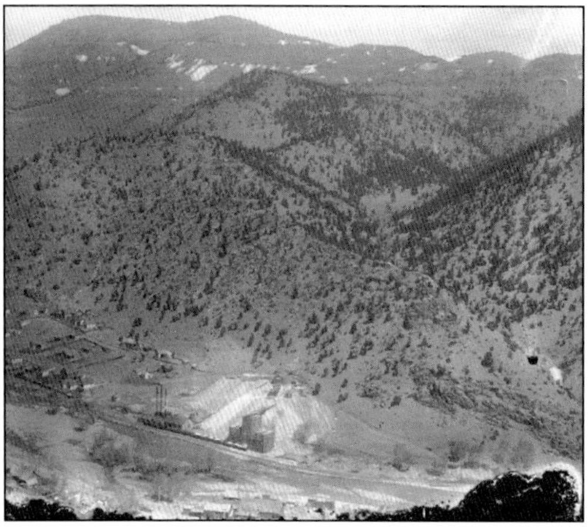

Here is a view of the renamed Argo Tunnel in 1906. The tunnel was a success almost overnight. While most of the attached companies had contracts to drain their mines, one did not. It paid for the Argo under the table. Following the closure of the gold mines for World War II, the old mines that were on the books were accessed and drained, and the one not on the books was backed up for several years, causing a great lake of water to form underground. Miners working the tunnel in 1946 struck the underground reservoir, causing a massive flood. The water pressure was so great that it shot all the way across the valley. (UCB.)

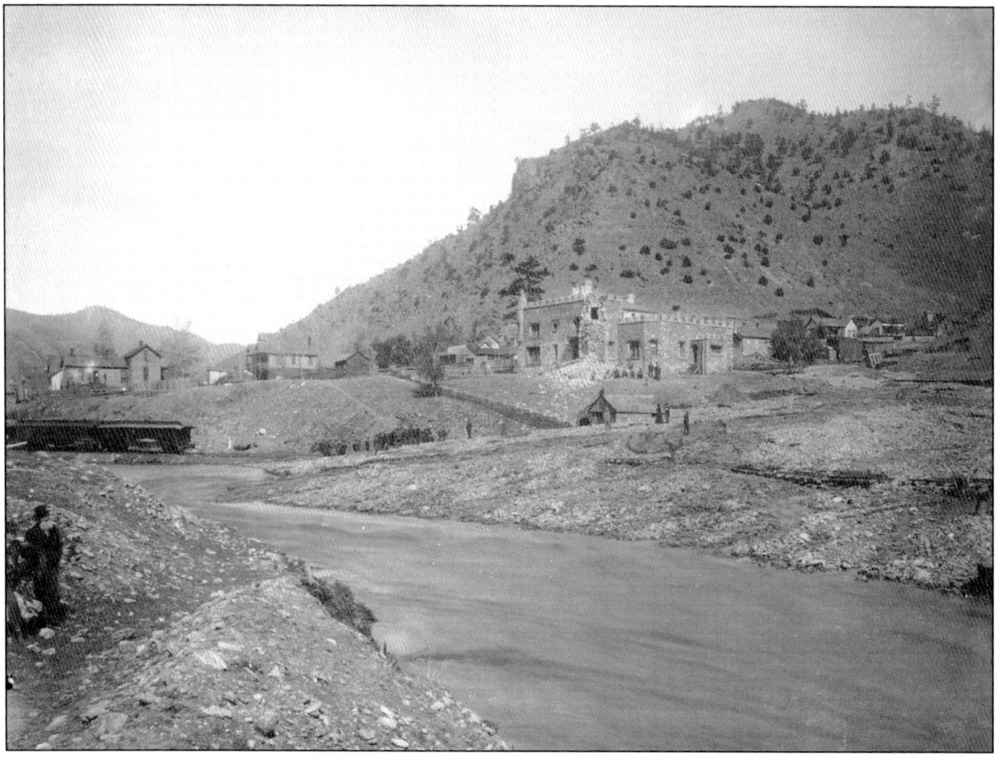

Castle Eyrie, built in 1882, stands right center, and the railroad runs along the creek. The castle is situated at the mouth of Virginia Canyon. (UCB.)

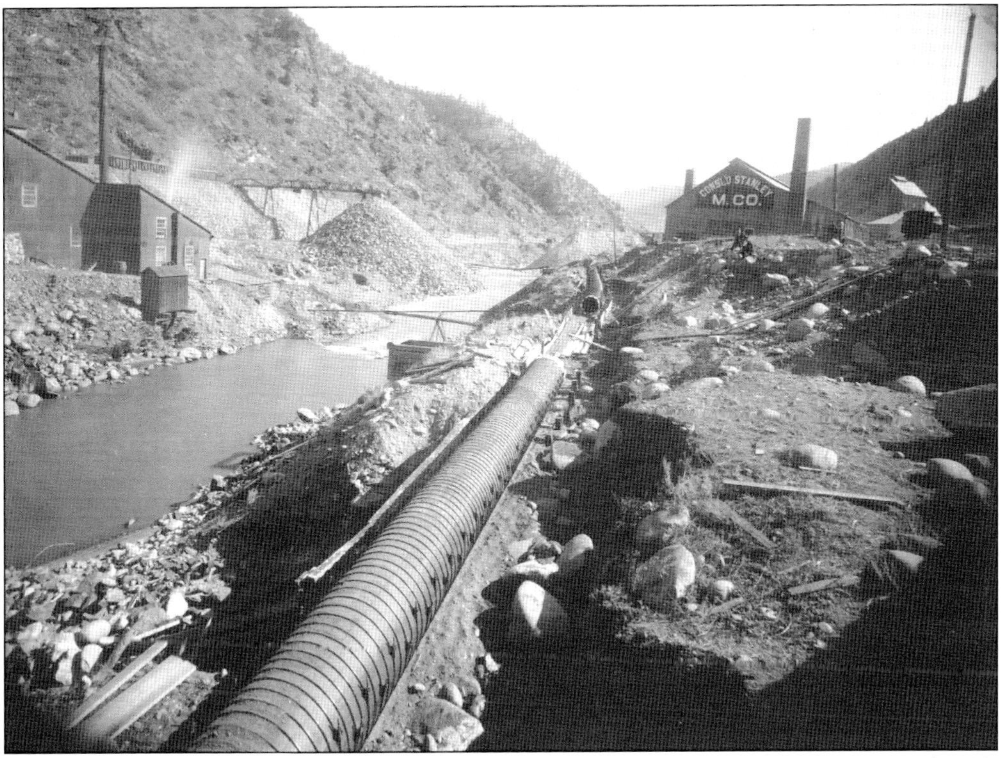

Here is the installation of a water flume at the Consolidated Stanley Mining Co. around 1903. The mill, as well as the powerhouse at the right, was eventually replaced with a new mill. There were 6,000 feet of underground workings at the Stanley. (UCB.)

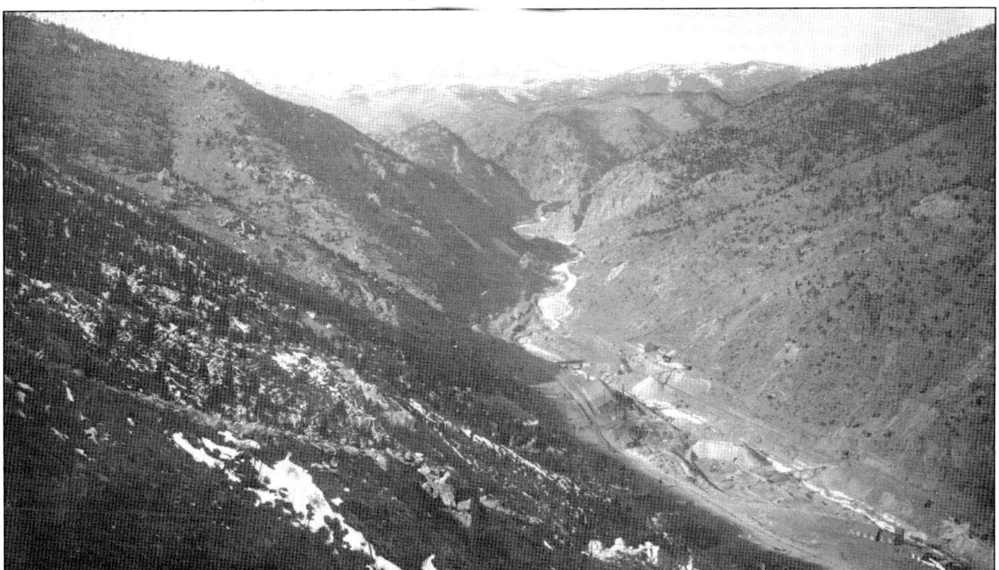

Seen from above Spanish Bar are the Stanley, Plutus, and Whale mining operations. The mouth of Fall River is at right center and is the only known place where a river runs into a creek. Clear Creek is seen running at center to lower right. (UCB.)

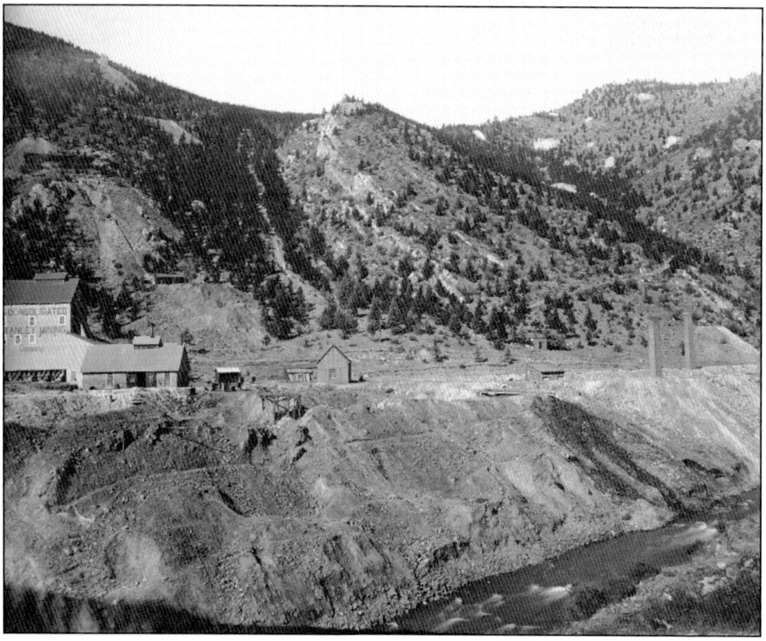

The Consolidated Stanley Mining Co. has expanded its operations up the hillside and has removed its former powerhouse and crushing equipment. (UCB.)

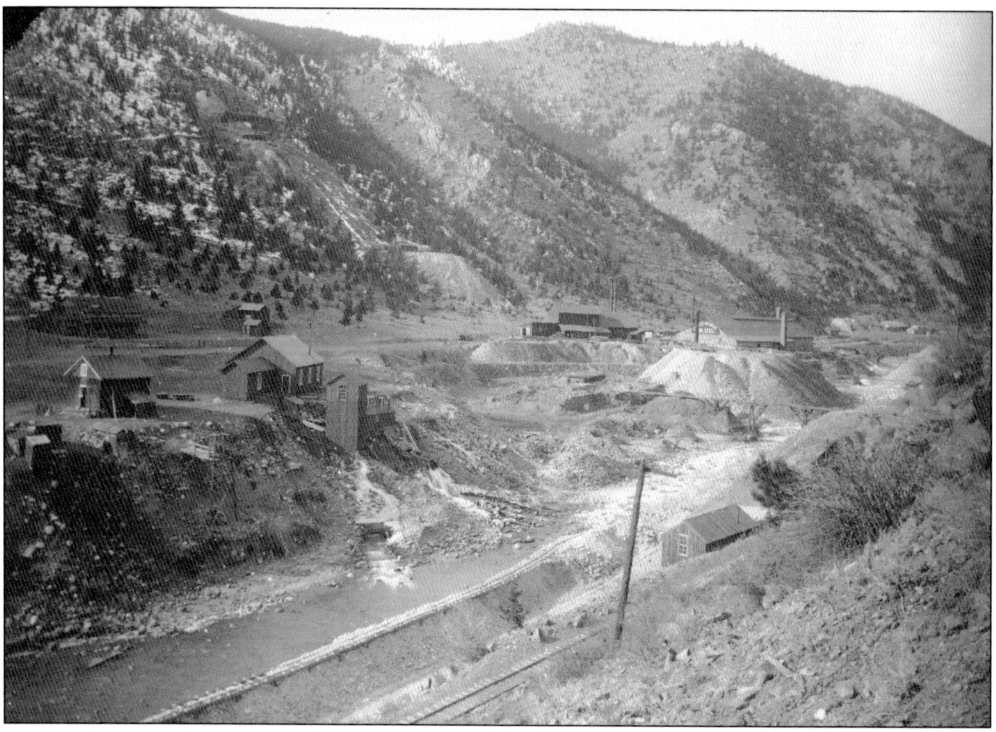

Railroad tracks of the Colorado & Southern, which took over the line from Union Pacific in 1899, follow Clear Creek up the valley. The original Stanley Mining Powerhouse and Mill are at right center. The shaft house and hoist works for the Plutus stands behind. The bridge crossing the creek at right center serves the mines of Hukill Gulch. Hukill Gulch was discovered in the 1870s but was not fully developed until bought by John Dumont, who also had mining claims at Empire and up the creek near Mill City (later renamed Dumont). (UCB.)

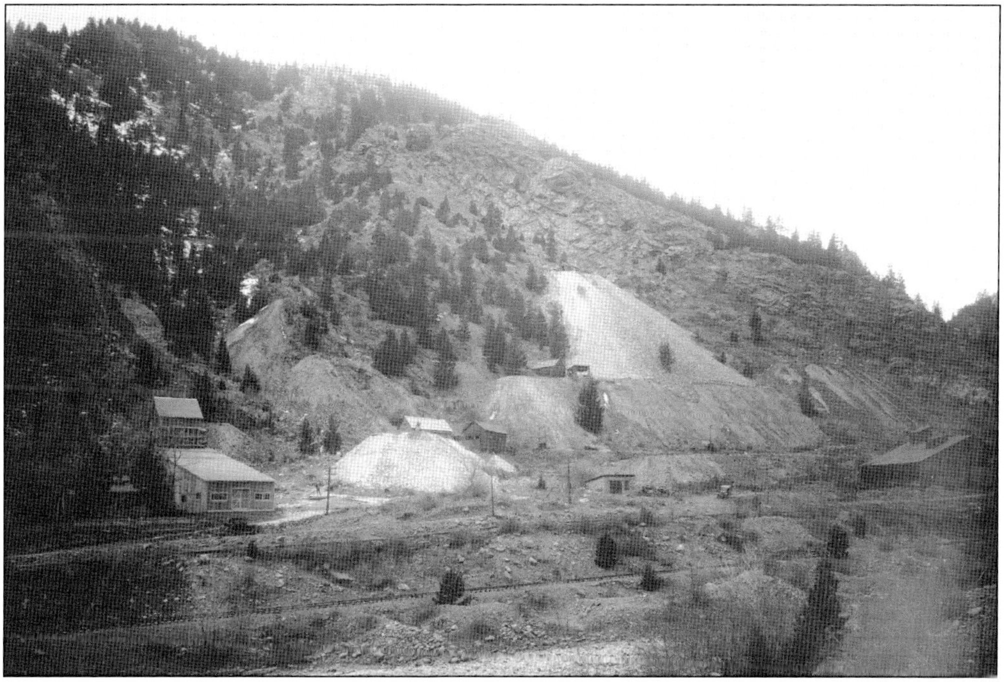

The Alma Lincoln Mine, discovered in 1863, is seen with its mill to the right of center. After a fury of activity on Spanish Bar, where Alma Lincoln is located, many placer miners moved on to Empire to cast their lots there. When they returned to Spanish Bar, many were surprised that their placer claims had now become the lode claims of the Alma Lincoln, Plutus, Whale, and Stanley. The three vertically climbing dump piles of the Elliott and Barber Mine are seen left of center, the Josephine Mine is right of center and above the stone powder house, and Lincoln Mine and Mill are at far right. By 1940, when this photograph was taken, the mill at right center was long abandoned. (DPL.)

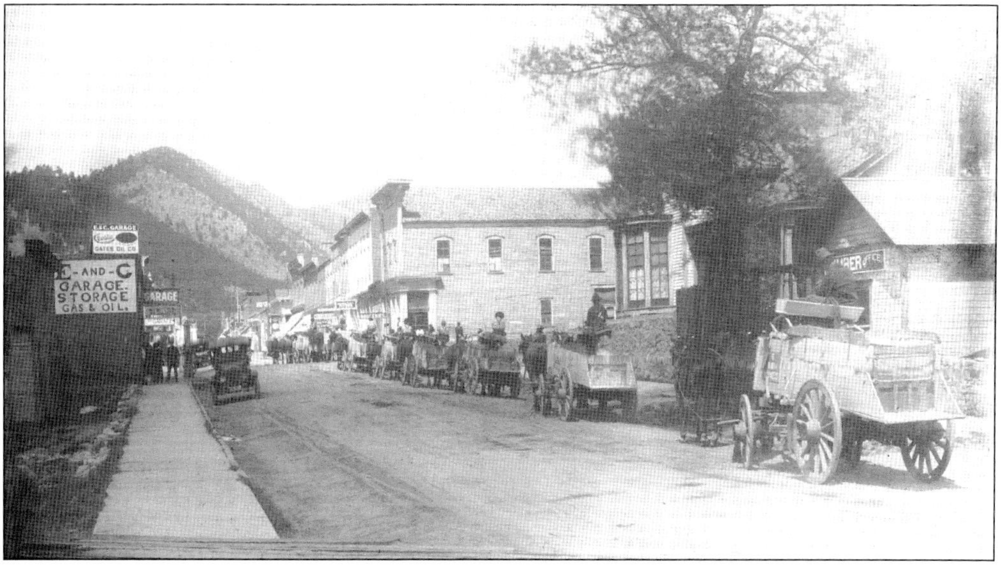

Gem ore wagons are along Miner Street in Idaho Springs around 1910. The railroad crosses the street, and so does a new form of transportation—the automobile. (UCB.)

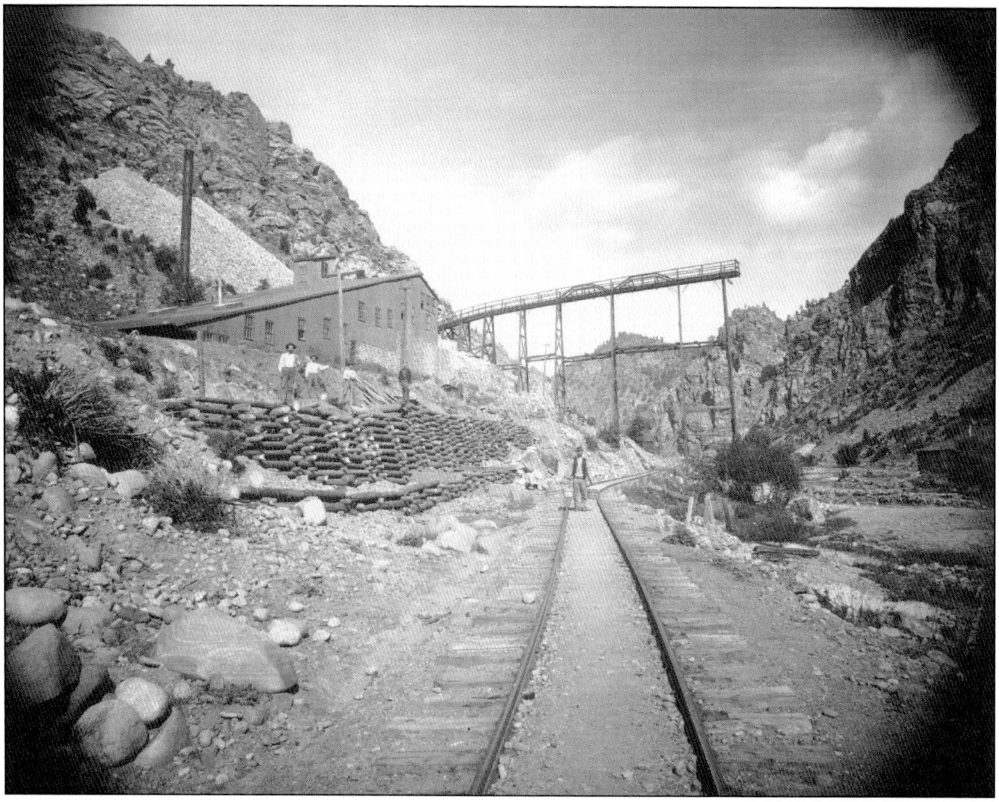

At its peak, the Alma Lincoln workings had become so extensive that it was necessary to build the high trestle to dispose of waste rock. Fall River's entrance is to the right beyond the trestle. Alice, Yankee Hill, Ninety-Four, and St. Mary's Glacier lie up the Fall River valley. (UCB.)

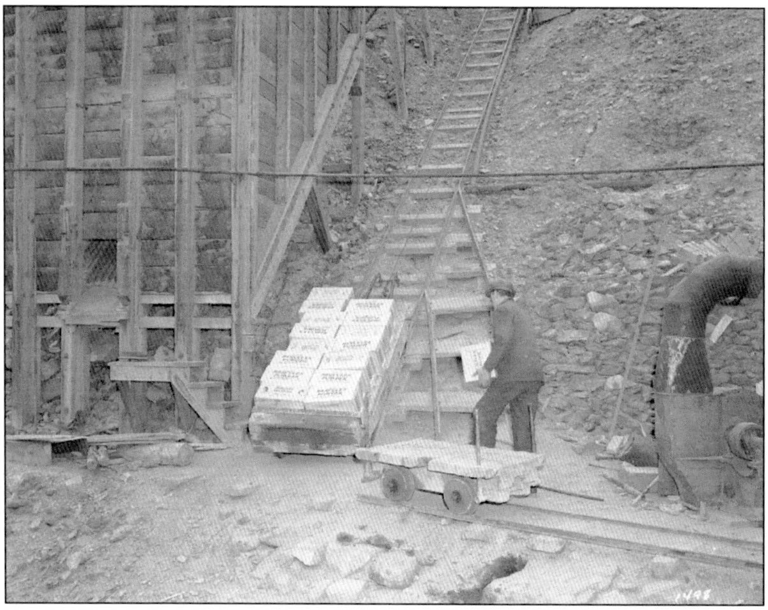

This worker is loading giant (dynamite) powder at the Alma Lincoln. The ore bin to the left is a classic example of the method used to load the ore into cars on the rails. Giant powder came to Clear Creek County in 1870; Alfred Nobel patented dynamite in Germany in 1867. (DPL.)

Hoosac Tunnel mill was located near the mouth of Fall River. The Hoosac Tunnel is a railroad tunnel built in Western Massachusetts in 1854. A homesick Eastern individual familiar with this area named the new Hoosac Tunnel in Clear Creek County. (UCB.)

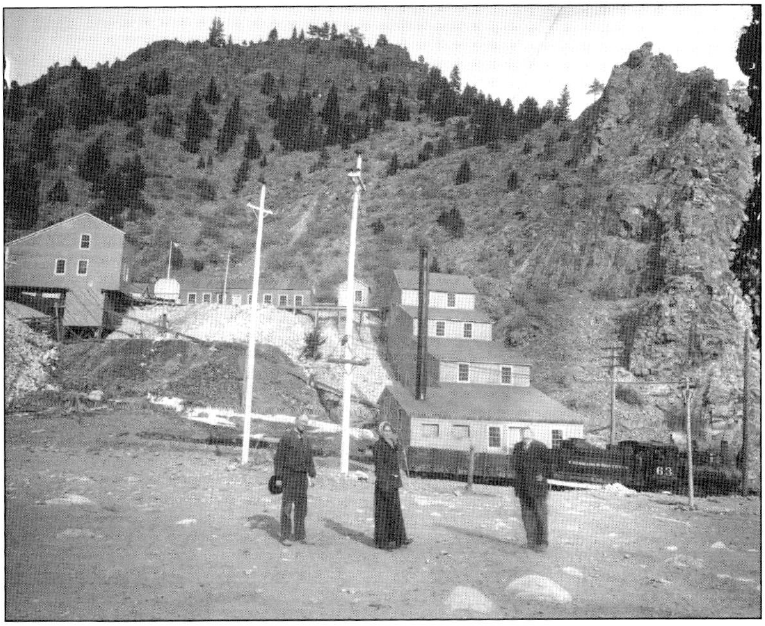

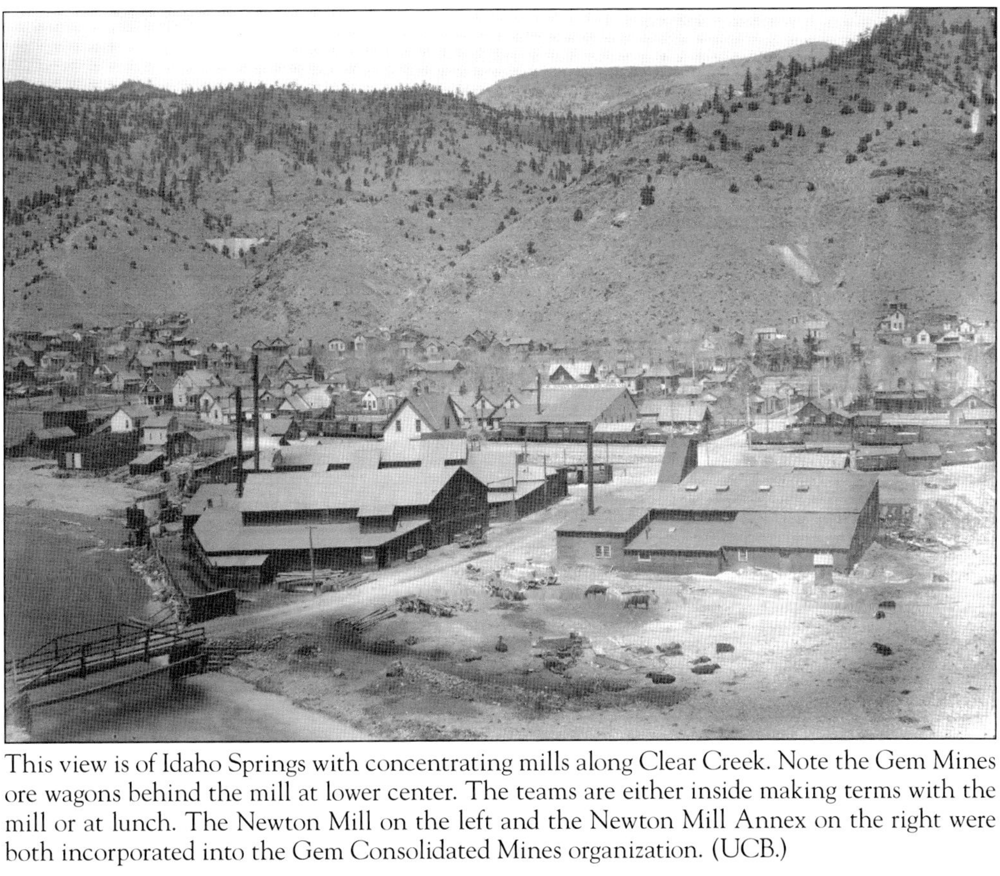

This view is of Idaho Springs with concentrating mills along Clear Creek. Note the Gem Mines ore wagons behind the mill at lower center. The teams are either inside making terms with the mill or at lunch. The Newton Mill on the left and the Newton Mill Annex on the right were both incorporated into the Gem Consolidated Mines organization. (UCB.)

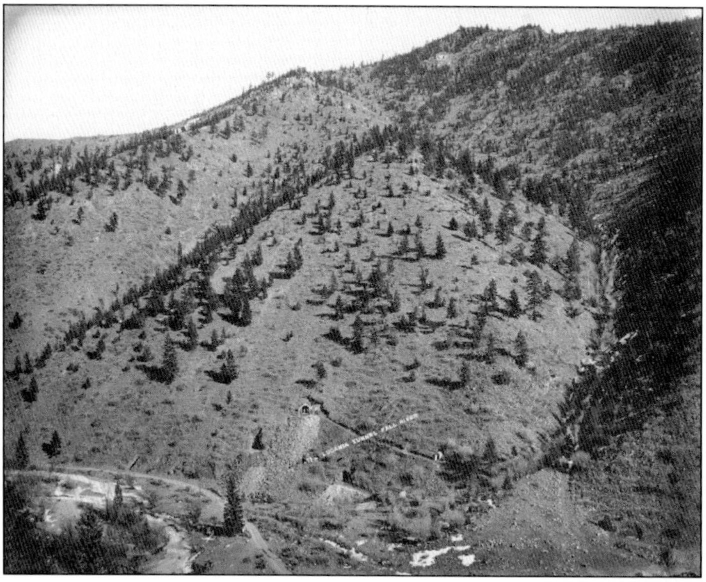

Here is an early view of Lucania Tunnel up Fall River. By 1917, the crosscut tunnel was driven to 6,240 feet from a point on the east side of Fall River Road. The project also connected with the Big Five Tunnel, located on the west end of Idaho Springs (see page 25.) (UCB.)

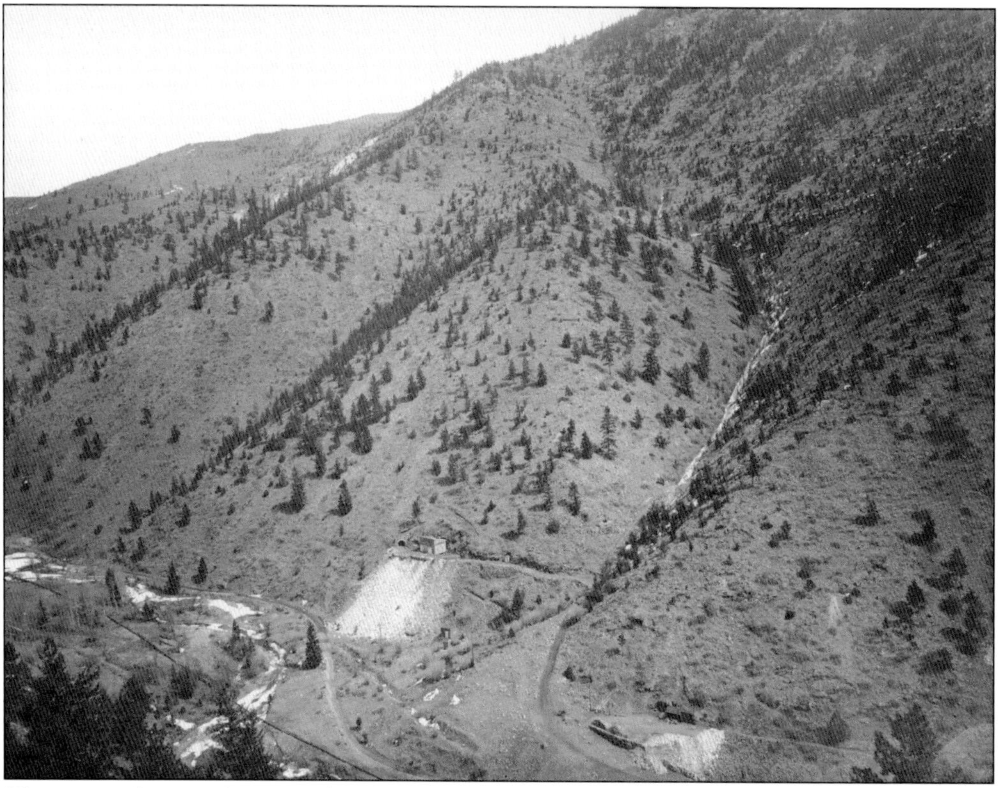

The waste pile grows bigger as the Lucania Tunnel progresses. The tunnel was funded by British investors and was named for the RMS *Lucania*, a Cunard ocean liner from Liverpool that set the Atlantic crossing speed record in 1894. The tunnel was designed to run all the way to Russell Gulch and Quartz hill in Gilpin County, to cut those mines at considerable depths, and to reach the Gilpin County Patch above Central City about 14,800 feet from the portal at a depth of 1,400 feet, but never made it. (UCB.)

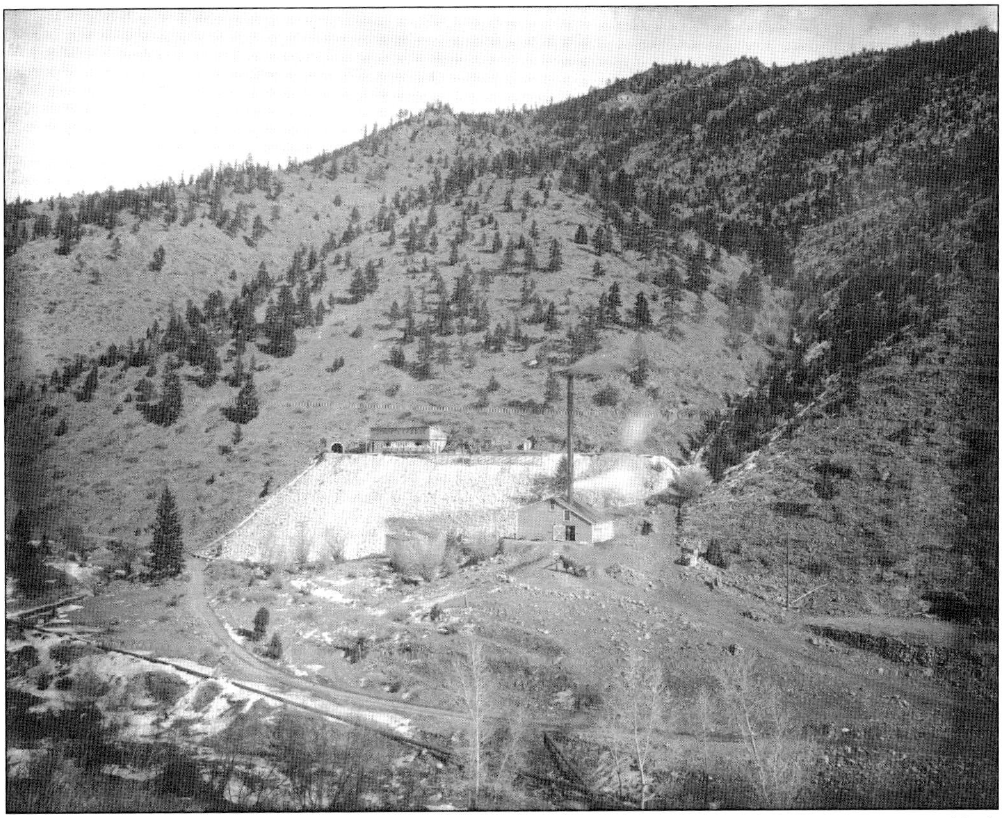

As the Lucania Tunnel progressed and more investment money came in, a powerhouse was added. By this time, pneumatic air drills were being used instead of the old single- and double-jacking methods to drive the tunnel. (UCB.)

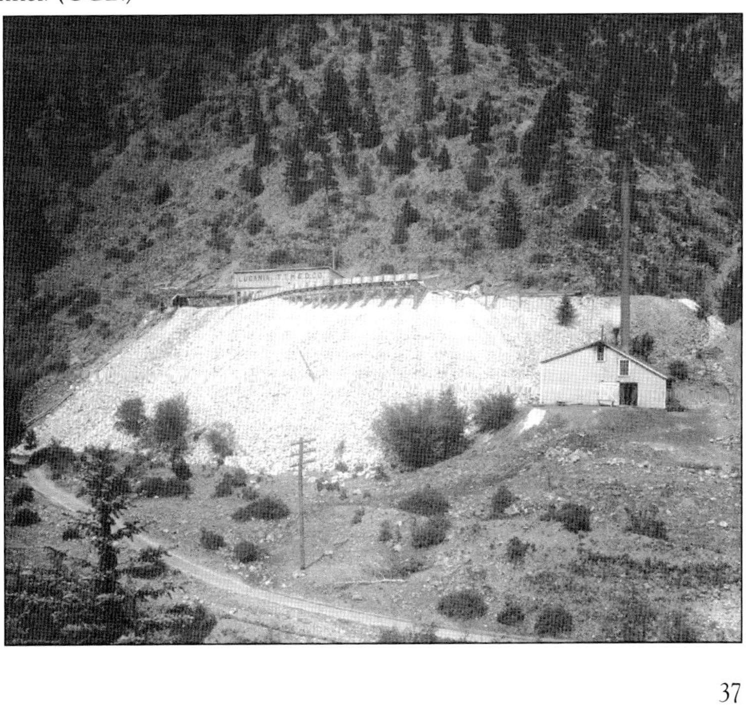

This view is of the Lucania Tunnel. By this time, workers have added a building and have piles of piping to place air ducts to pump in fresh air. (UCB.)

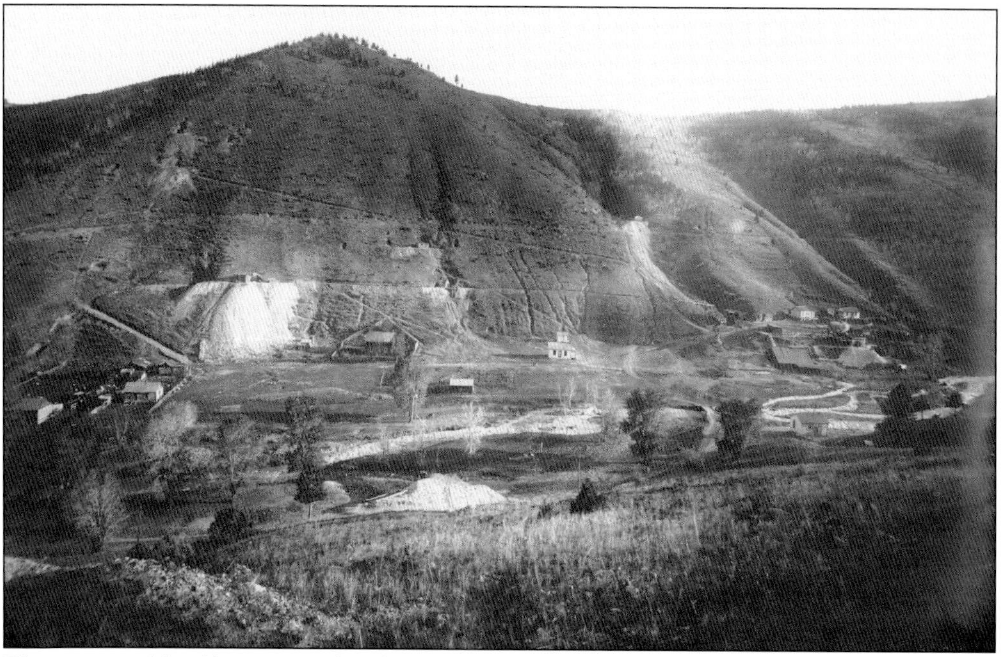

This view is of the Chicago Creek School and Egan Reduction Works on Chicago Creek, located up from Idaho Springs. George Jackson's gold discovery in January 1859 was at the confluence of Chicago Creek and Clear Creek in Idaho Springs. (UCB.)

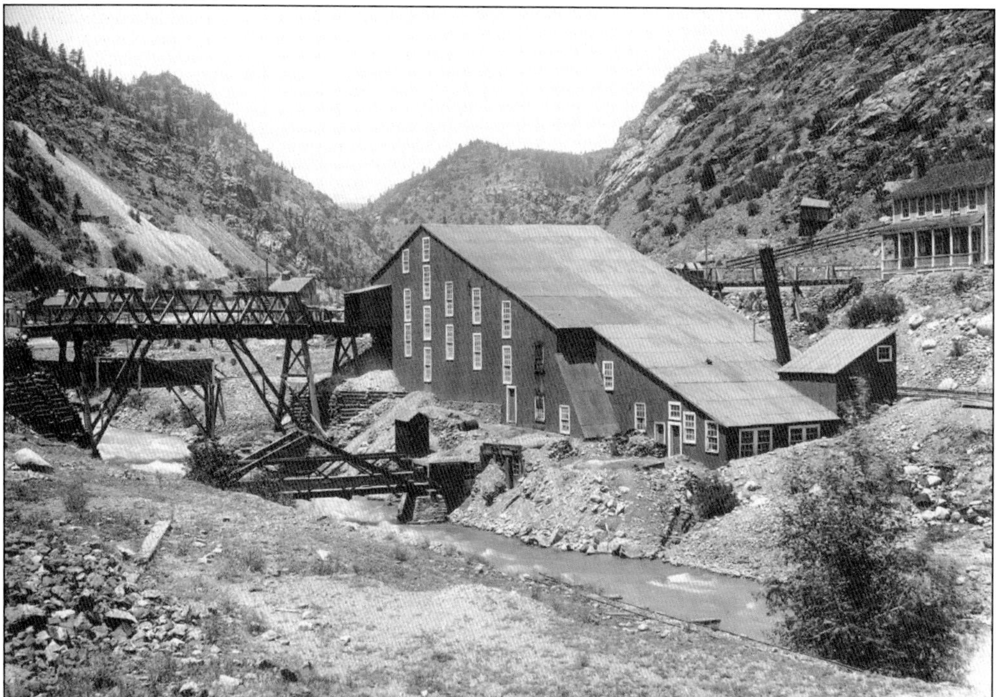

This view is of the Plutus Mine and Mill on Spanish Bar near Idaho Springs. The manager's house is seen at right. A close look at the Alma Lincoln Mine (left center) shows the 75-foot-tall trestle coming out of the mountain. (UCB.)

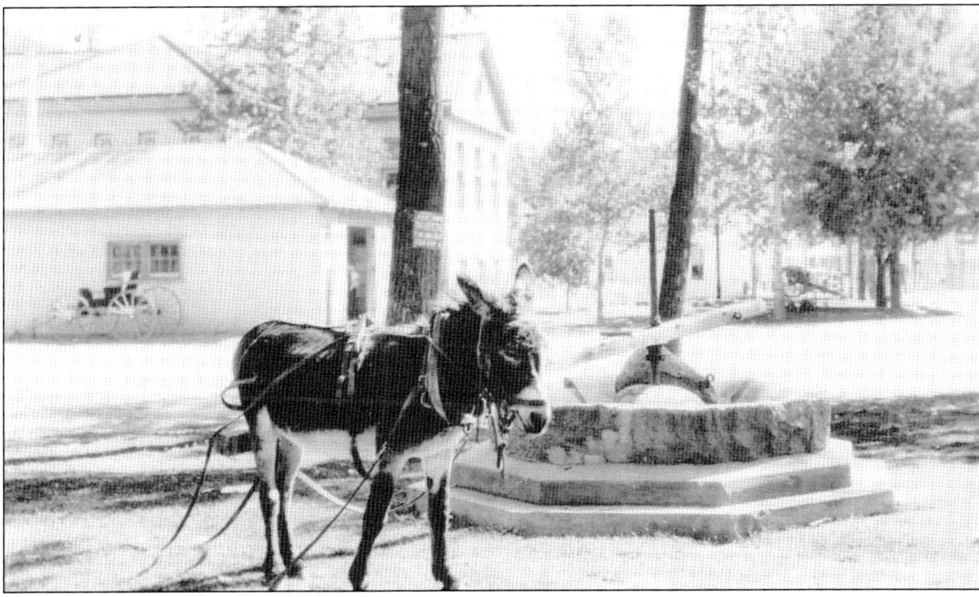

Located in Idaho Springs, this Spanish arrastra was used to crush rock at Dumont. Spanish Bar was named for the early Spanish mining methods found there. Still visible today, there is a square tunnel opening high above the Stanley Mine indicative of the Spanish mining methods. This photograph was taken after 1913. A donkey was hooked up to the arrastra to show people how it worked. It was later moved next to the Carnegie library at the old city hall, pictured left center. Note the fire department building between the carriage and city hall. This structure replaced the fire building and bell tower (see page 19) after the opera house and building were torn down in 1913. Many old-timers were sad to see the old structure go. (DPL.)

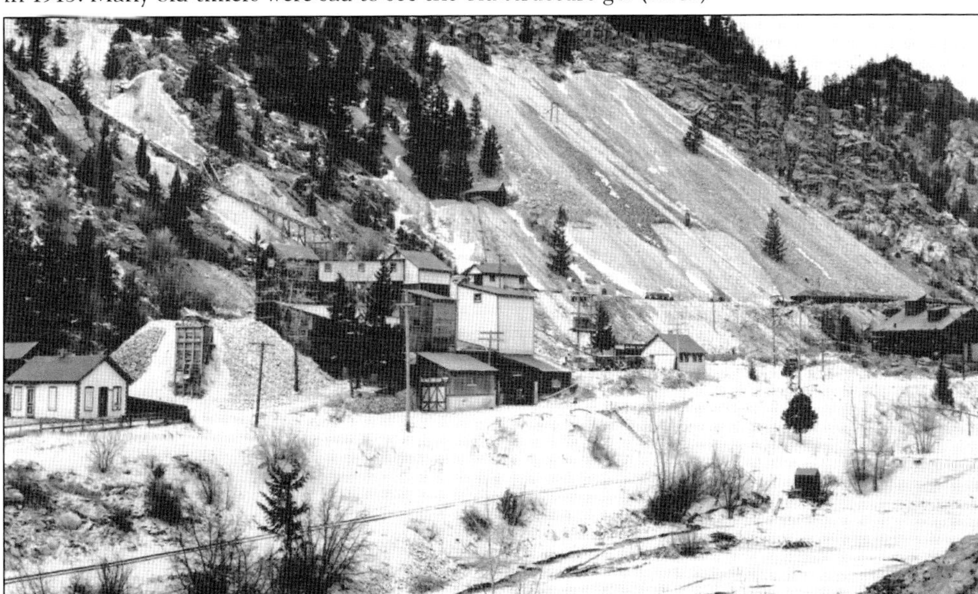

The Alma, Lincoln, Elliott, and Barber Mines are pictured in 1940. Note the mill and trestle from the image on page 34 are gone. The mill burned and the trestle was removed. The mill (right center) lies abandoned after many years. (DPL.)

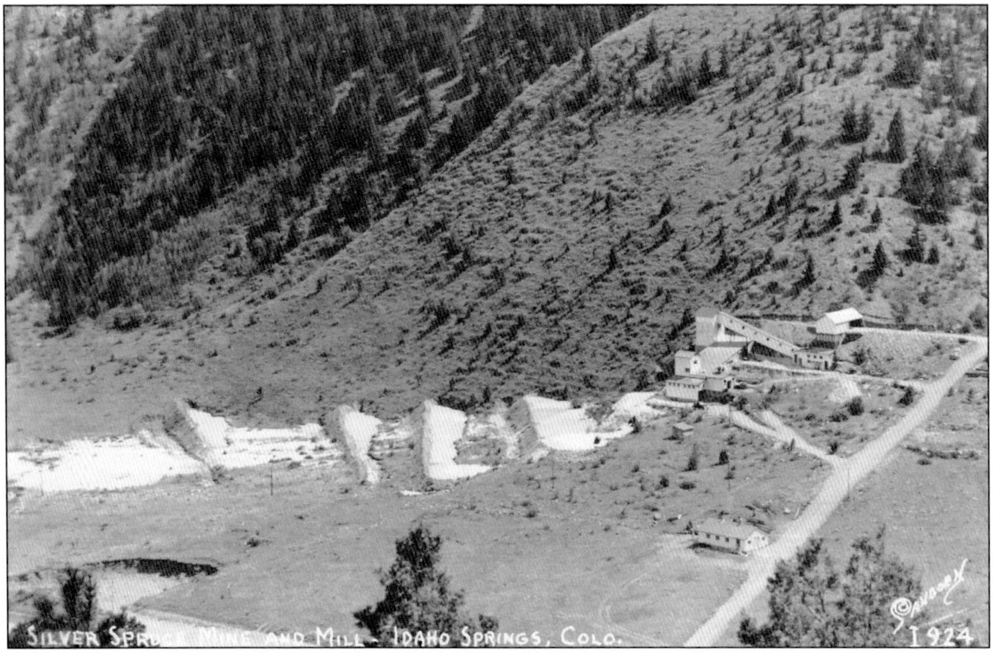

This view is of the Silver Spruce Mine and Mill at the east end of Idaho Springs in the 1930s. At that time, the price of gold went from $20 an ounce to $35 an ounce, making it possible to open many old abandoned mines.

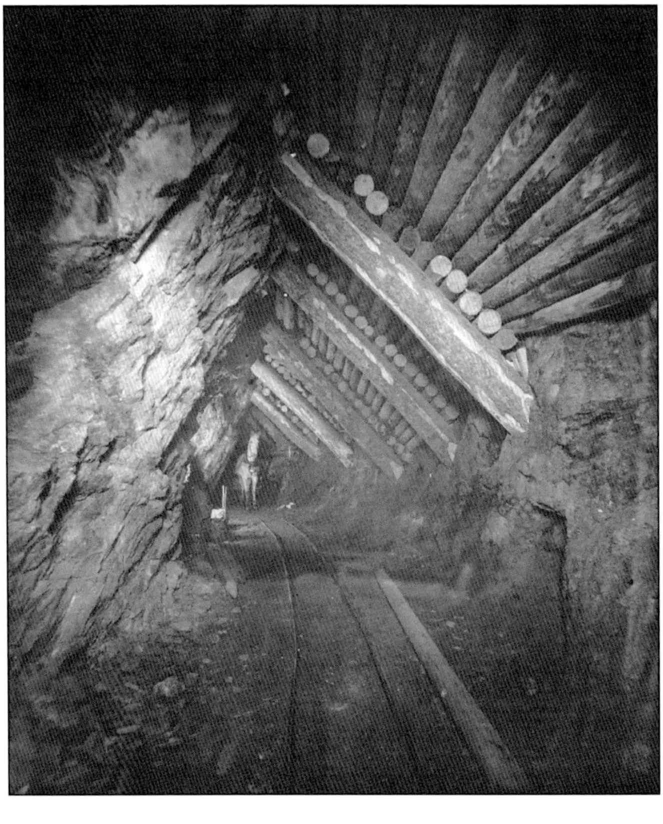

Here is an interior underground view of the Stanley Mine workings. This is a donkey hauling ore cars with a timbered stope above. This was a common method of elevating the tunnel. As the vertical workings climbed higher, timbers would be placed above the main access tunnel to protect the passageway from debris above.

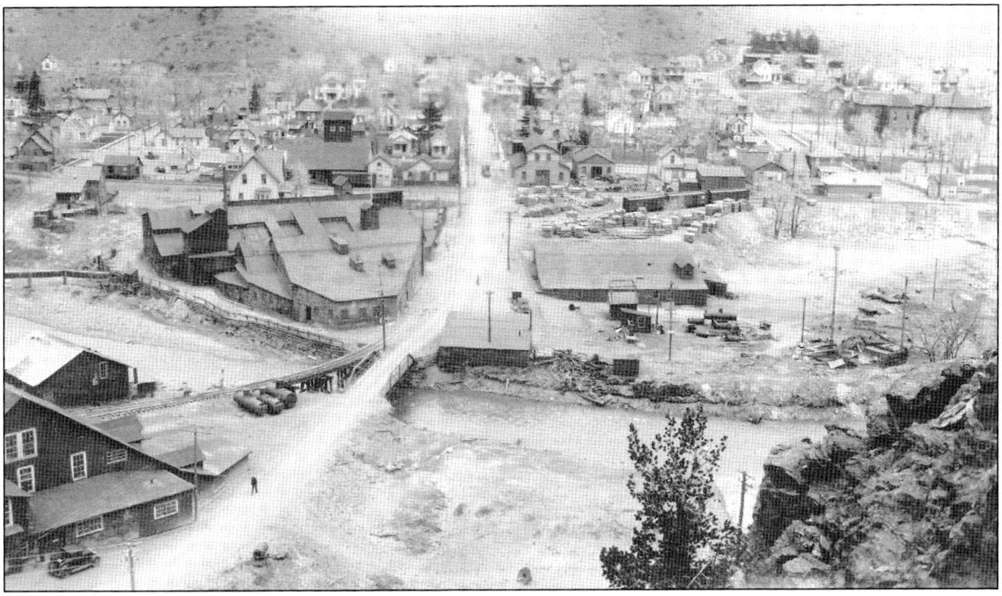

Pictured are the abandoned mills at the confluence of Clear and Chicago Creeks. Farther down Clear Creek, the Mixsell Dam and Mill were located. George Jackson first discovered placer gold (gold ore directly available from the stream) at this site in January 1859. Note the three boilers staged for removal near the bridge. (USGS.)

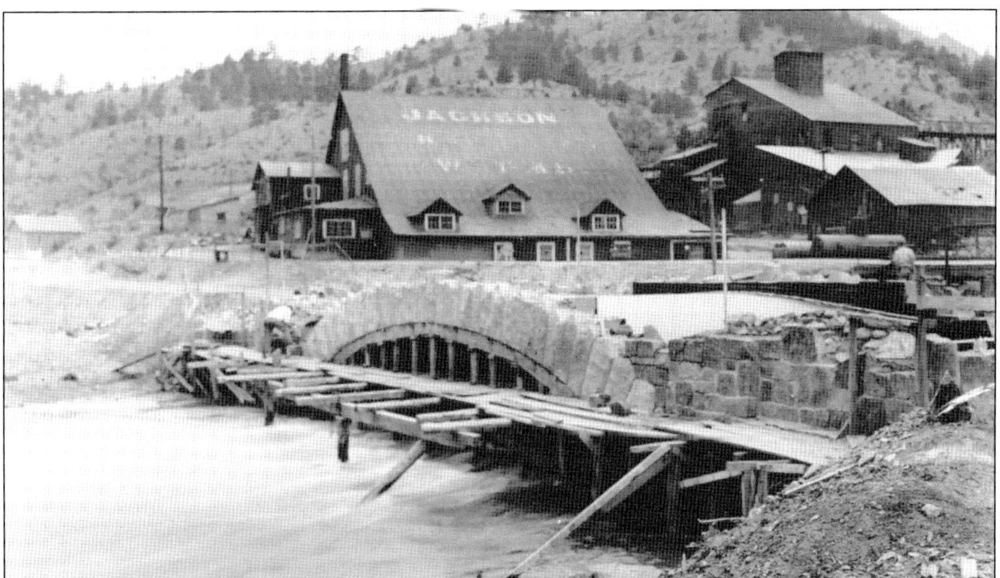

Bridge construction for the new and improved Mount Evans Highway was completed in 1931. Note old Jackson Concentrating works in the background and the Hudson Reduction Co. to the right. (USGS.)

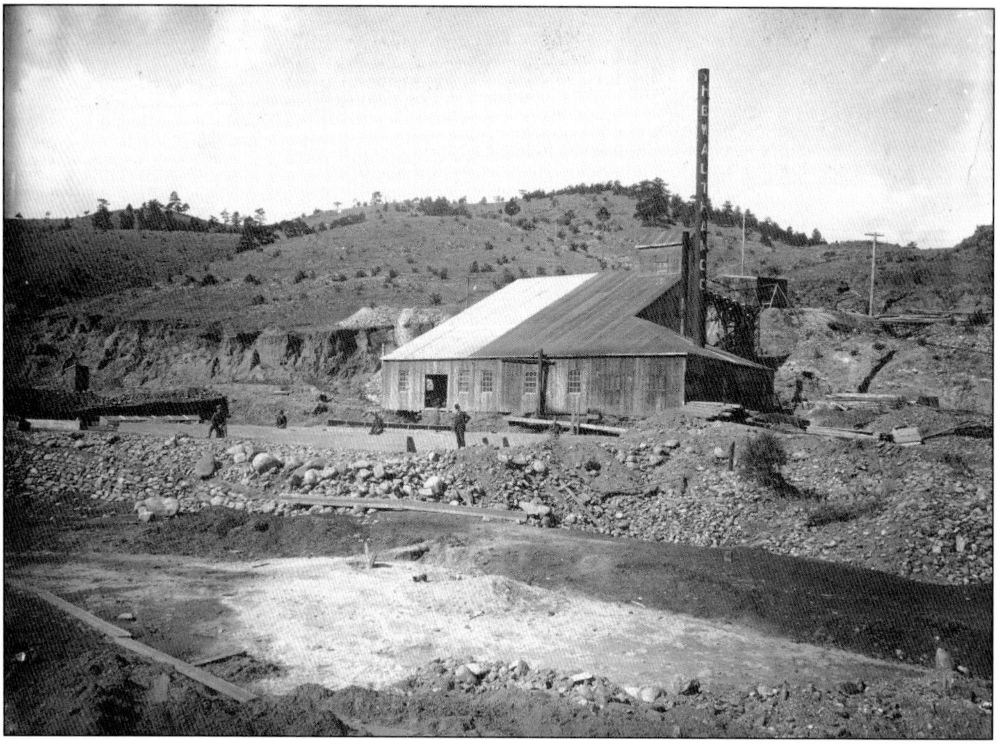

Pictured here is the Waltham Mill in Idaho Springs. The building in the photograph's center is the new potassium cyanide plant that was added in 1909. Using cyanide was an improved metallurgical technique for extracting gold from low-grade ore. (Both, UCB.)

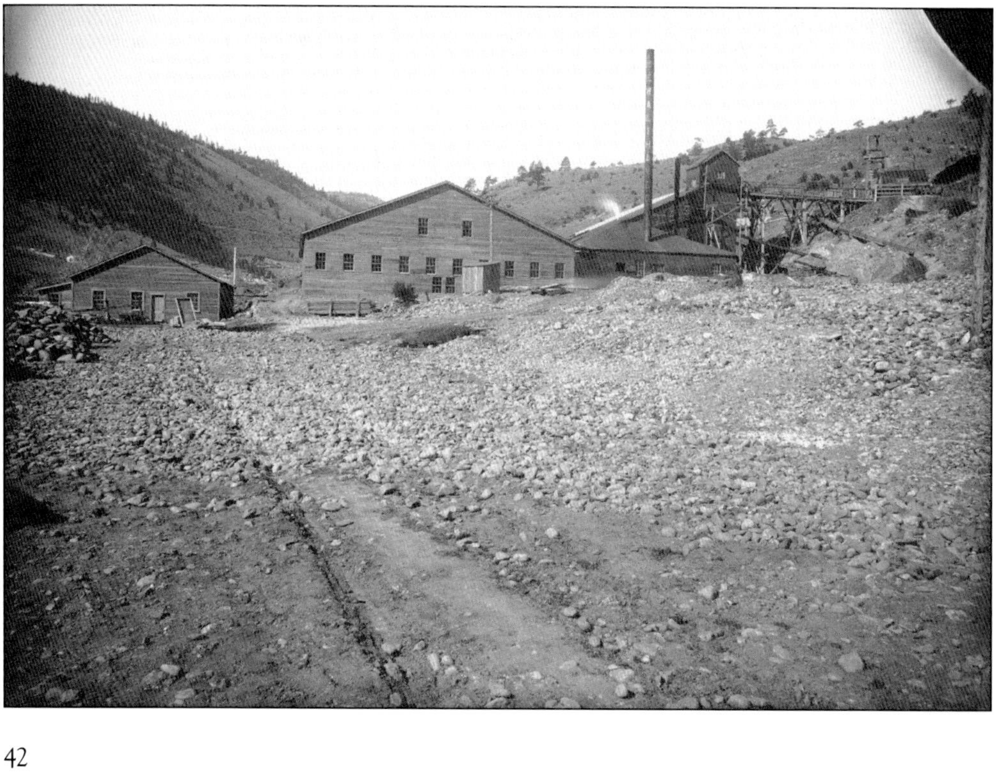

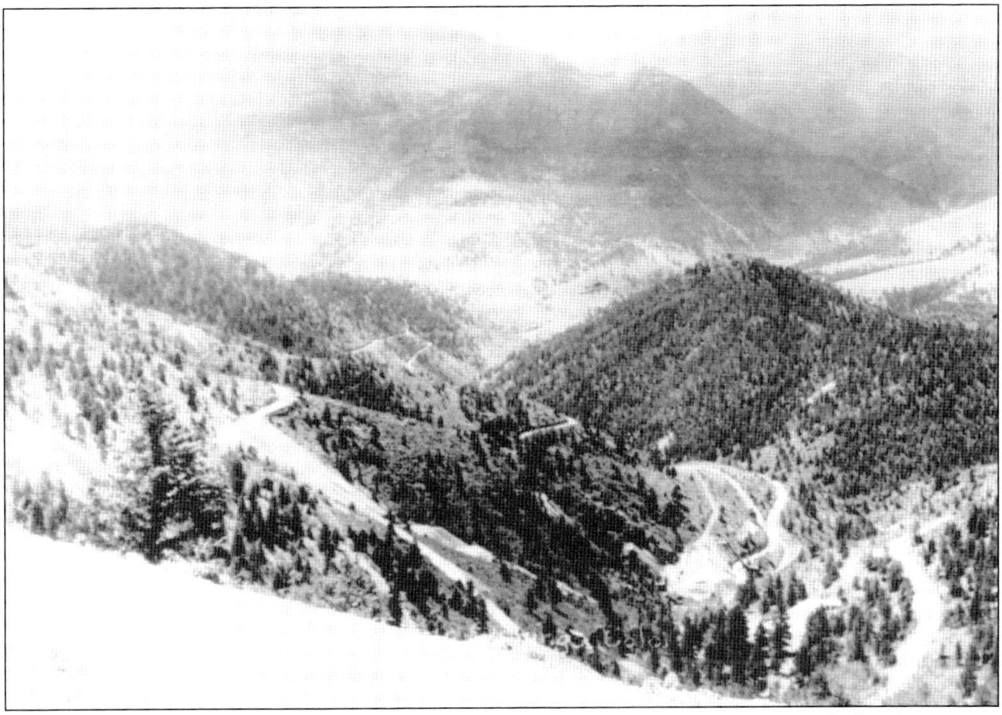

Here is Virginia Canyon seen from above Idaho Springs. Different roads were made for the wagons to haul the ore to the mills. They all connected to the main road from Central City to Idaho Springs, the Oh My God Stage Road.

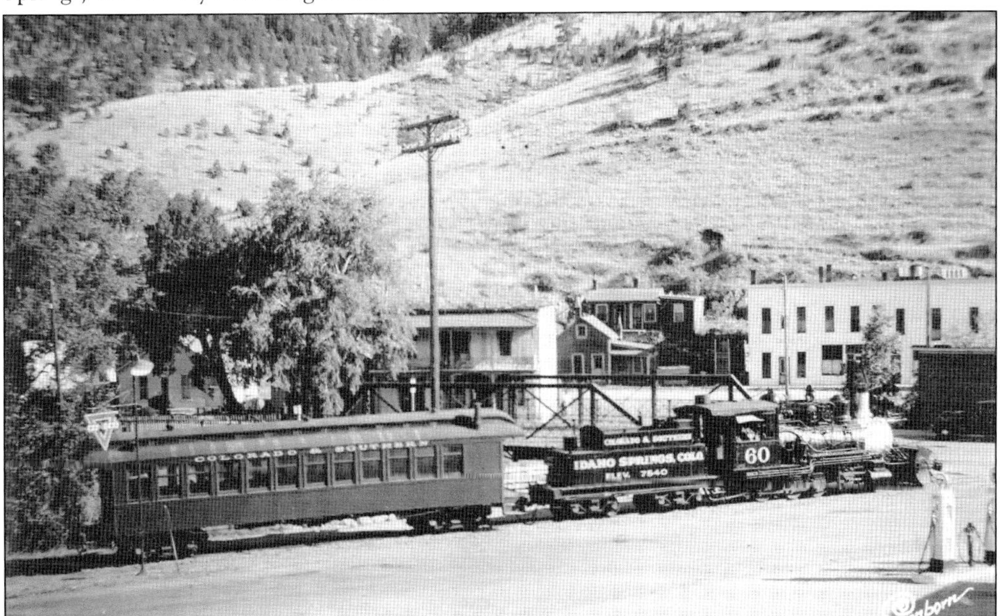

Old Smoke and Cinders, the Colorado & Southern train No. 60, was retired in place. The final train trip in Clear Creek County was in 1939. When the Grass Valley School was relocated from the Grass Valley Mining District to be used as the Idaho Springs Town Hall, the train was moved across Miner Street to become part of the town hall display complex.

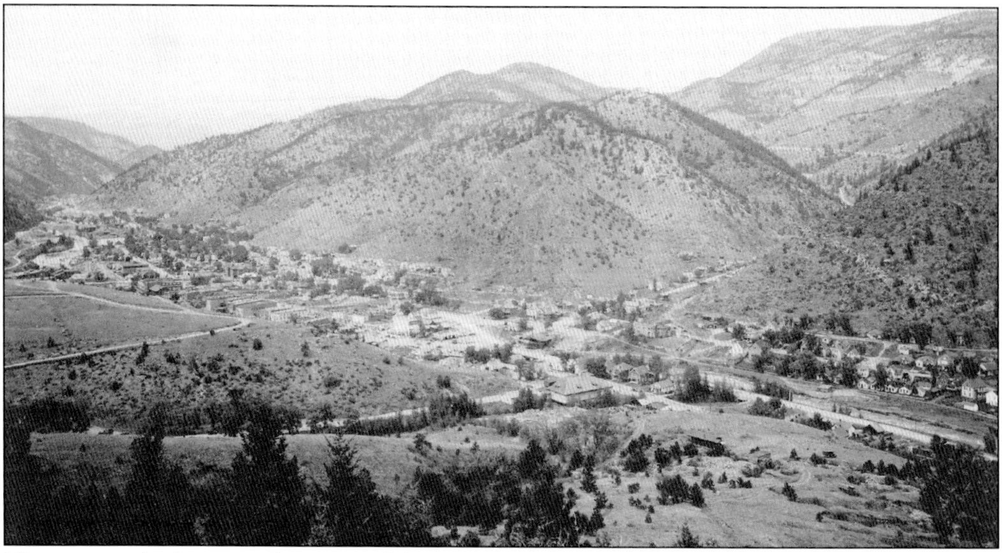

This view is looking northwest at Idaho Springs. The Trocadero Dance Emporium is located lower center at the mouth of Soda creek. Virginia Canyon goes off to the right over the top to Central City. (DPL.)

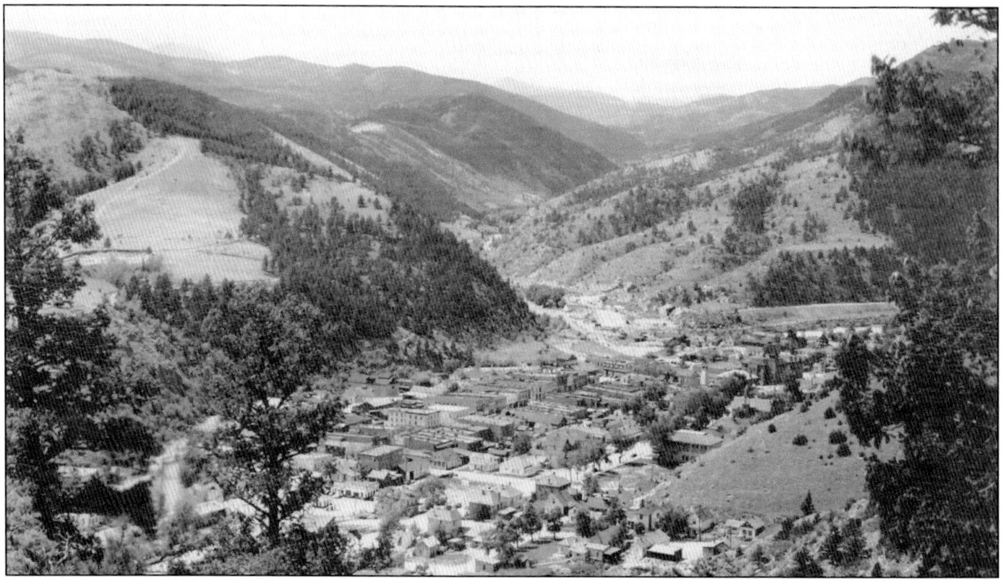

This view of Idaho Springs looks up the Chicago Creek (right center) around 1930. Note that the mills at the mouth of Chicago Creek are gone. A close look at right center shows the big brick school still standing. This school was condemned and torn down in the 1950s. By the late 1930s, all of the mills at the Jackson Discovery site had been dismantled.

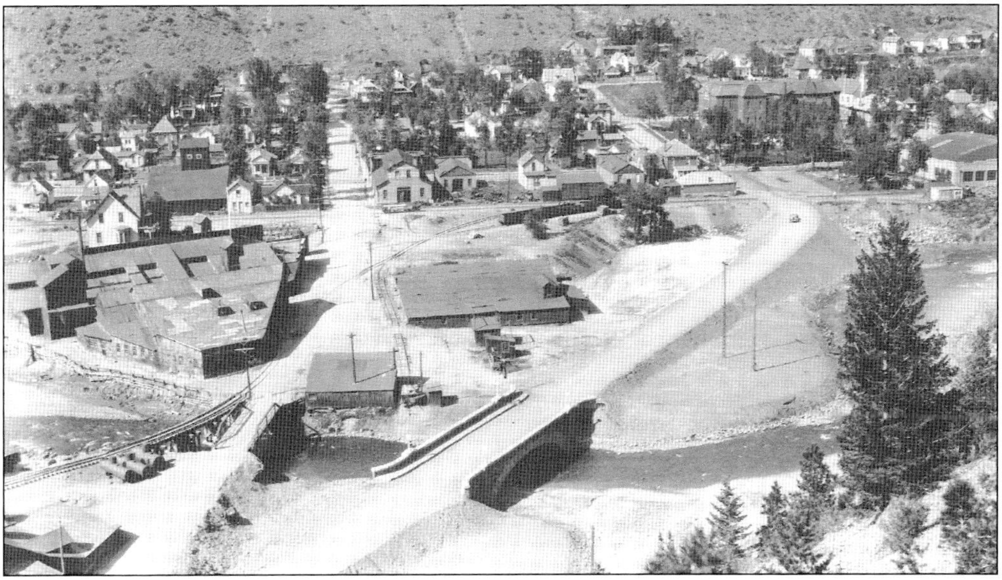

This view is of the newly constructed bridge at Idaho Springs in 1931. At this time, all of the mills are abandoned and being scrapped. Railroad service was still operating but not at a profit. It took the Colorado & Central Railroad several years to convince the transportation administration to discontinue service. The last railroad trip was in 1939. The old Kilton Reduction Sampling Works (upper left center) burned down in 1980. The fire was so hot in that it melted the red emergency light on the closest fire truck responding to the fire. (USGS.)

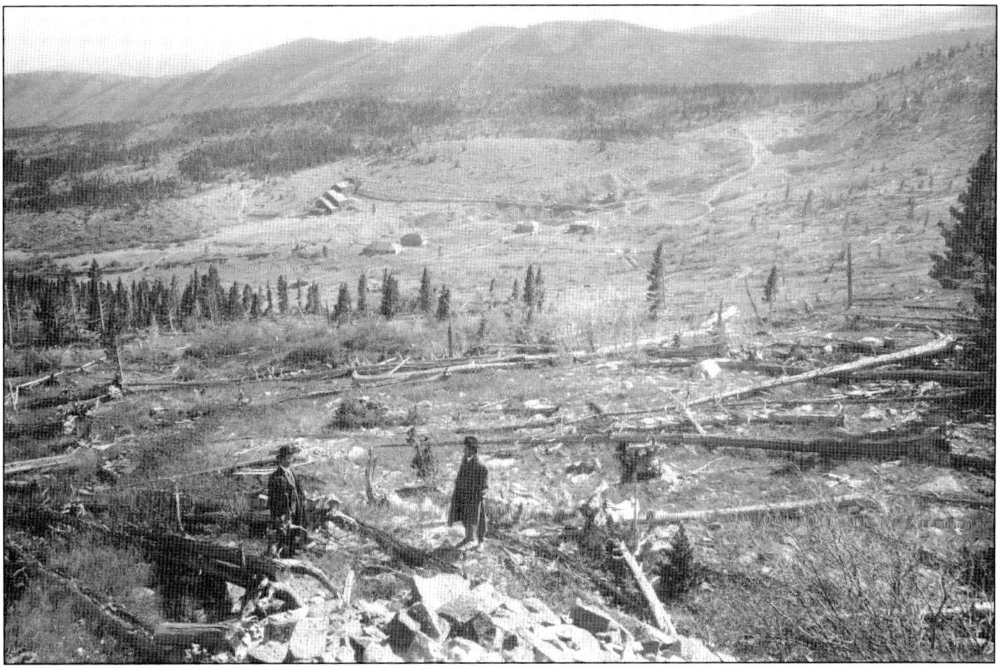

Here is a view of the Alice Mine and Glory Hole operation in 1911. This was first worked as a placer, producing 2,900 ounces of gold. Later, a glory hole was dropped, and a tunnel runs out of the bottom of the glory hole. Glory hole mining is a method of mining used to maximize the gold output of a certain mining area. (UCB.)

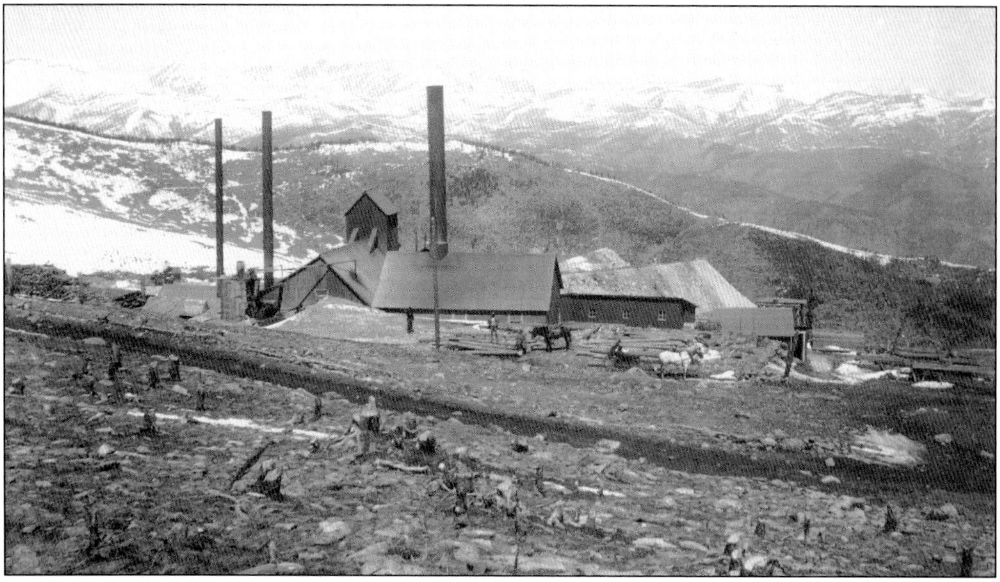

Here is the Lamartine Mine in a 1901 view looking northwest. The mine was discovered in the late 1860s but was not fully developed for 20 years, when it was managed by Lafayette Hanchett. It became a reliable property and a sizable producer, based on earnings of $616,000 over a period of 16 months. As of 1905, the total value of production was $2,361,039. Lamartine was named for Alphonse de Lamartine, a French writer, poet, and politician who died in 1869 at age 78. He is considered the first French romantic poet and was famous in his time for publishing *Cours familier de littérature*. Lamartine is located up above Chicago Creek along Ute Creek and is also accessible from Trail Creek near the mouth of Fall River. (UCB.)

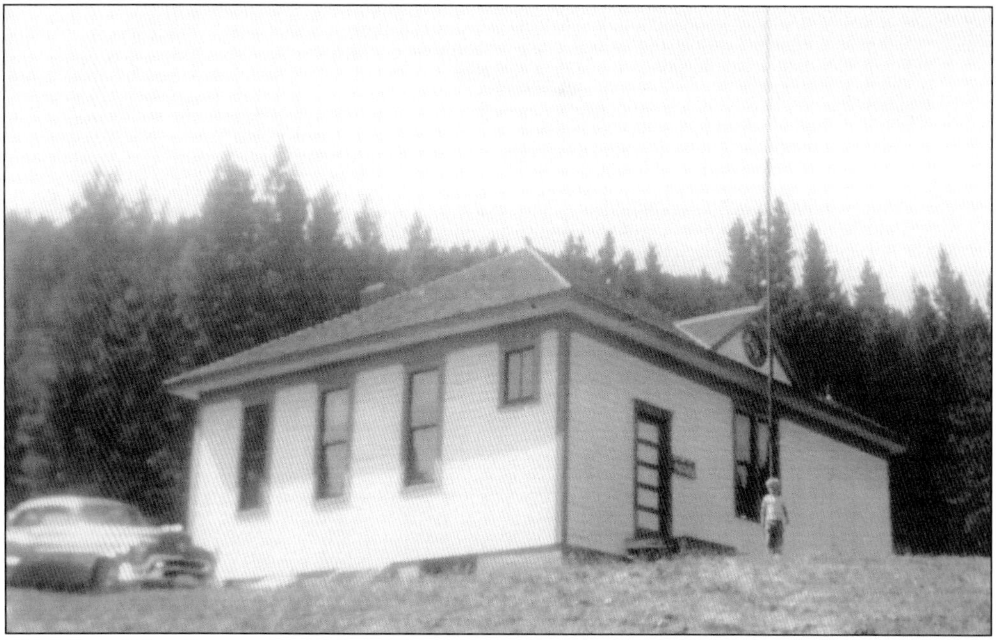

The Alice Schoolhouse was built in 1915 and operated until 1939. It is one of the remaining Alice structures still visible today. (DPL.)

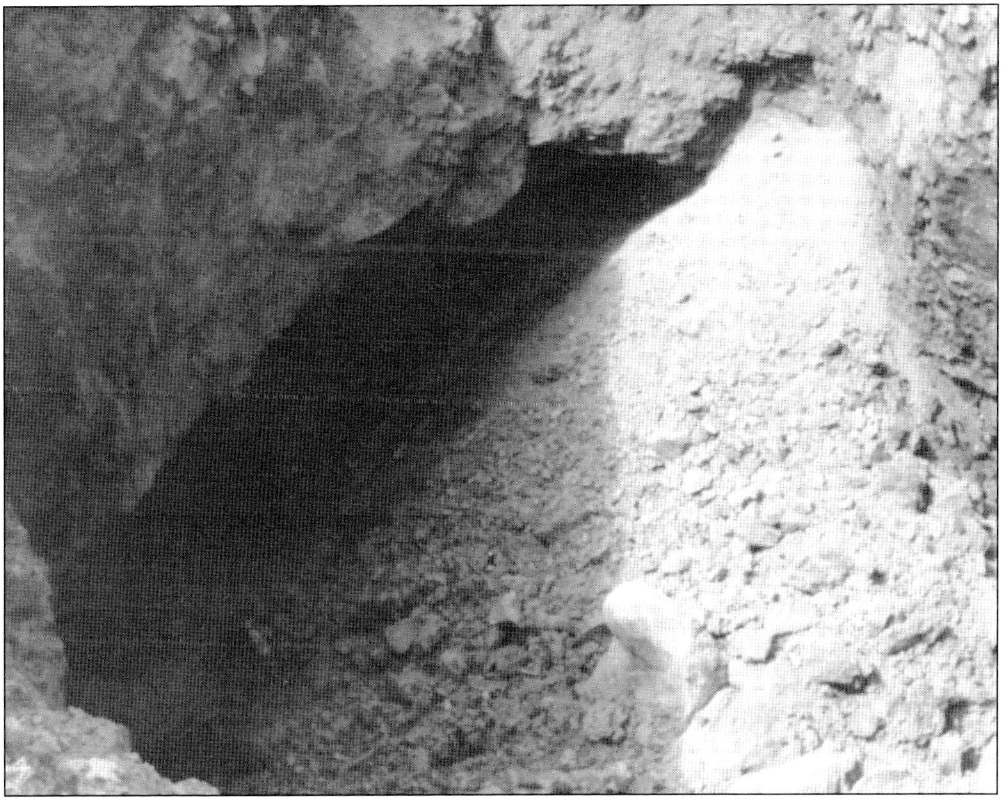

This interior view of the Alice Glory Hole dates to 1979. This hole was large and deep and was filled in by the EPA in the late 1980s. The glory hole was developed in the 1930s to maximize out the gold area. A similar glory hole was also developed for the mines in the Patch above Central City. The Chain O' Mines mined that area and had an aerial tramway until World War II deemed gold mines as nonessential and cut off the supply of explosives. Glory holing accesses the body of mineral from the bottom; the ore is then hauled out via a mine tunnel underneath the glory hole. (Both, Lynn Schimmelfeder.)

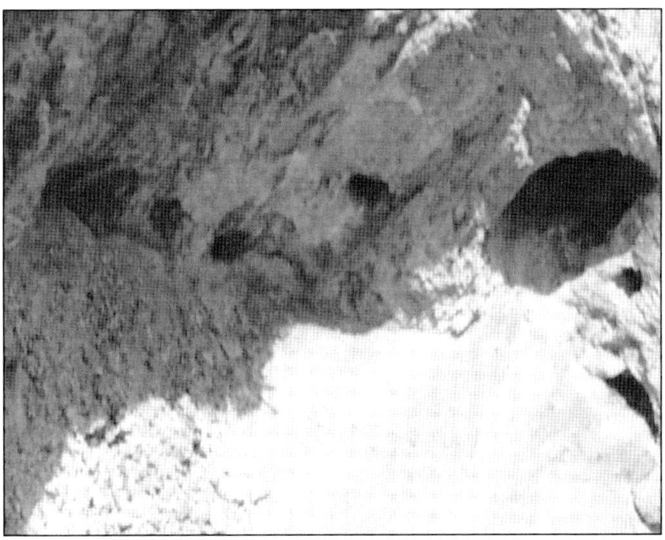

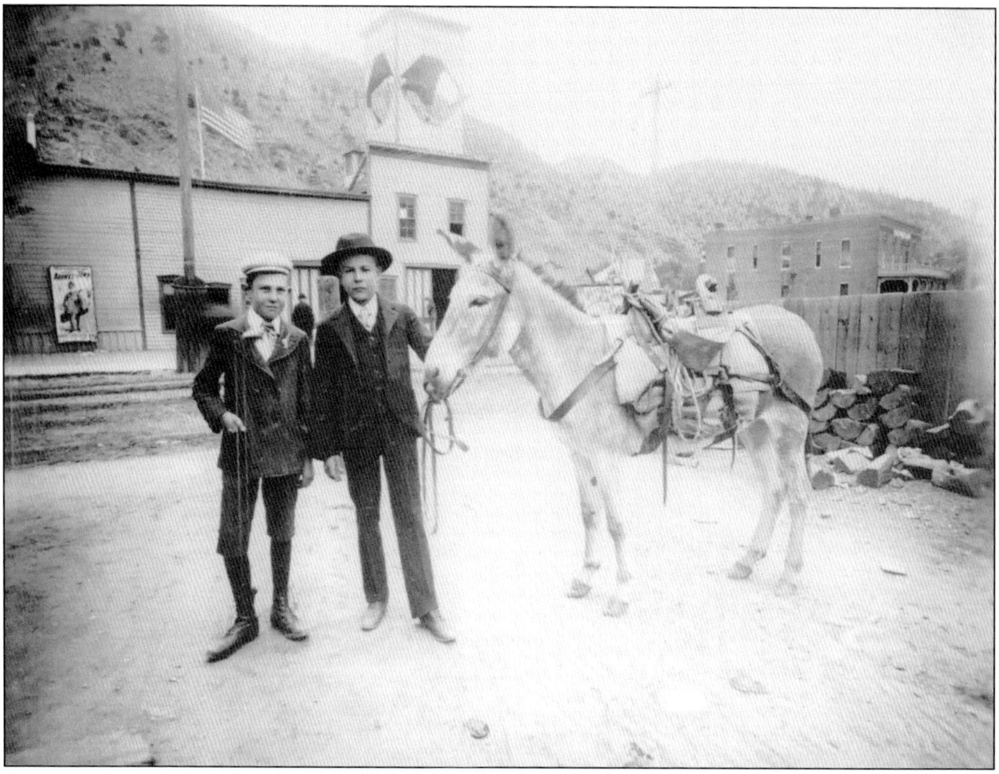

Here are two unidentified boys with a donkey in front of the Idaho Springs Opera house in the 1870s. The musical *Brown's in Town* is playing. With a bell in the tower, the fire department is visible behind the boys. The opera house was fairly sizable, as this Ancient Free and Accepted Masons dinner shows. The site was torn down in 1913. (Both, DPL.)

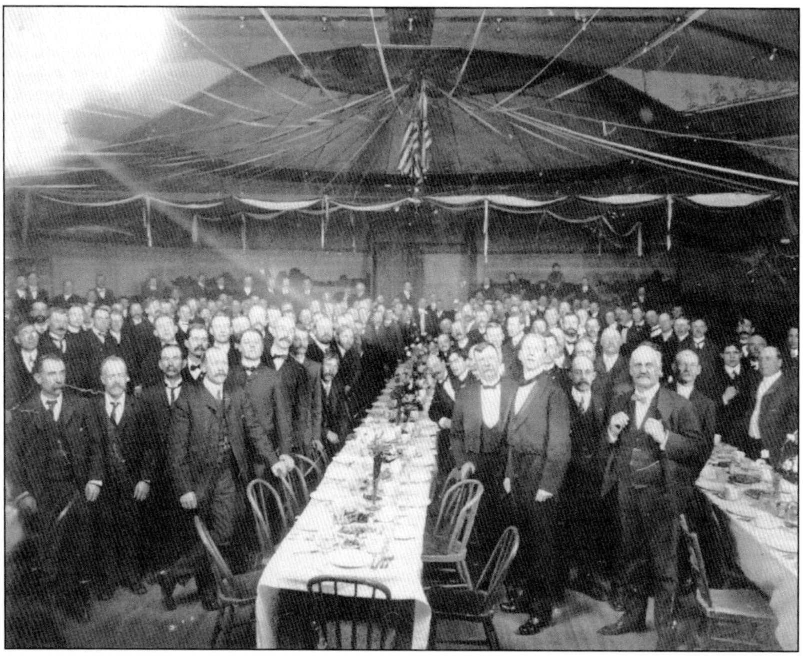

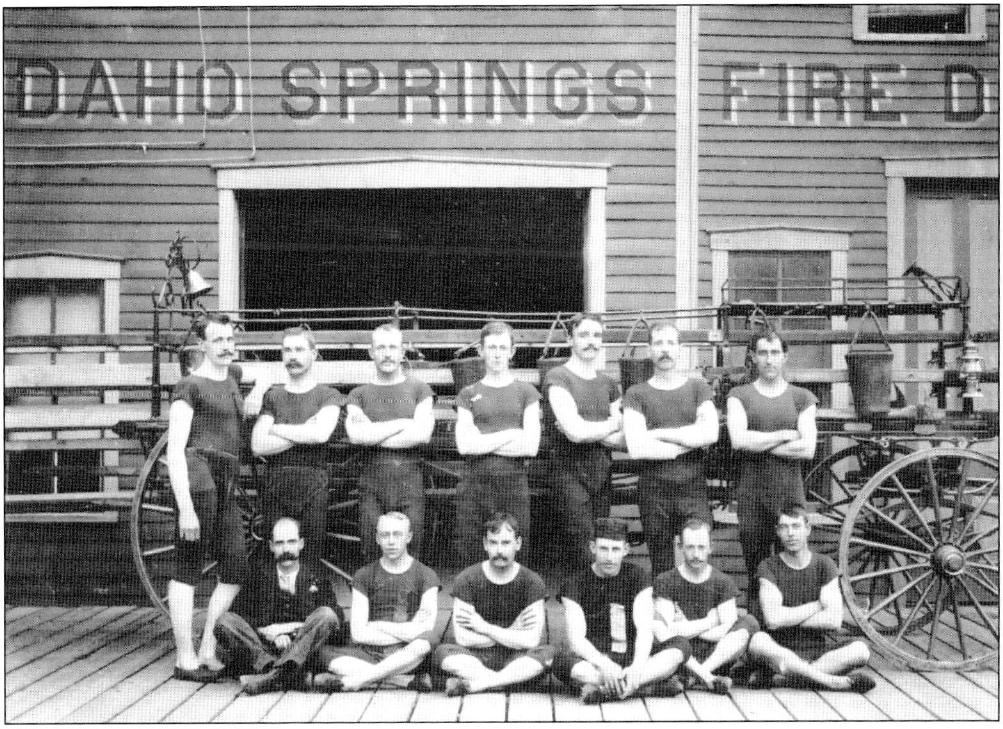

This view is of the Idaho Springs Fire Department crew dressed in full parade uniforms in 1895. These uniforms were also used for the hose cart races. They allowed full range of motion for the events. One had to be very athletic and active to be in the fire department. Strength and skill won hose cart races as well as put out fires. (DPL.)

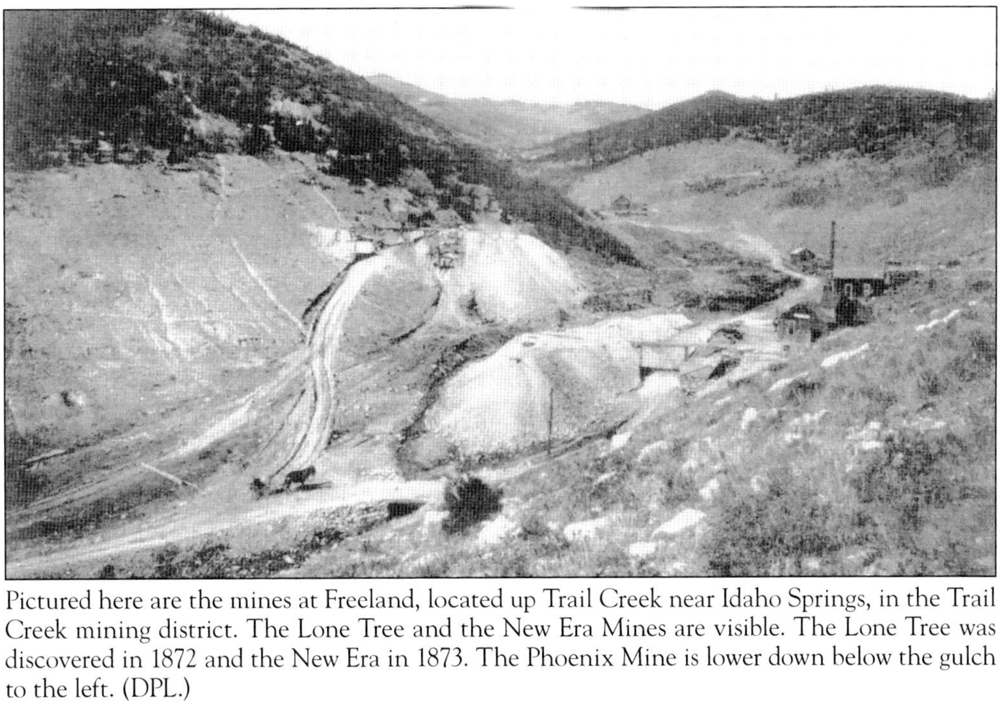

Pictured here are the mines at Freeland, located up Trail Creek near Idaho Springs, in the Trail Creek mining district. The Lone Tree and the New Era Mines are visible. The Lone Tree was discovered in 1872 and the New Era in 1873. The Phoenix Mine is lower down below the gulch to the left. (DPL.)

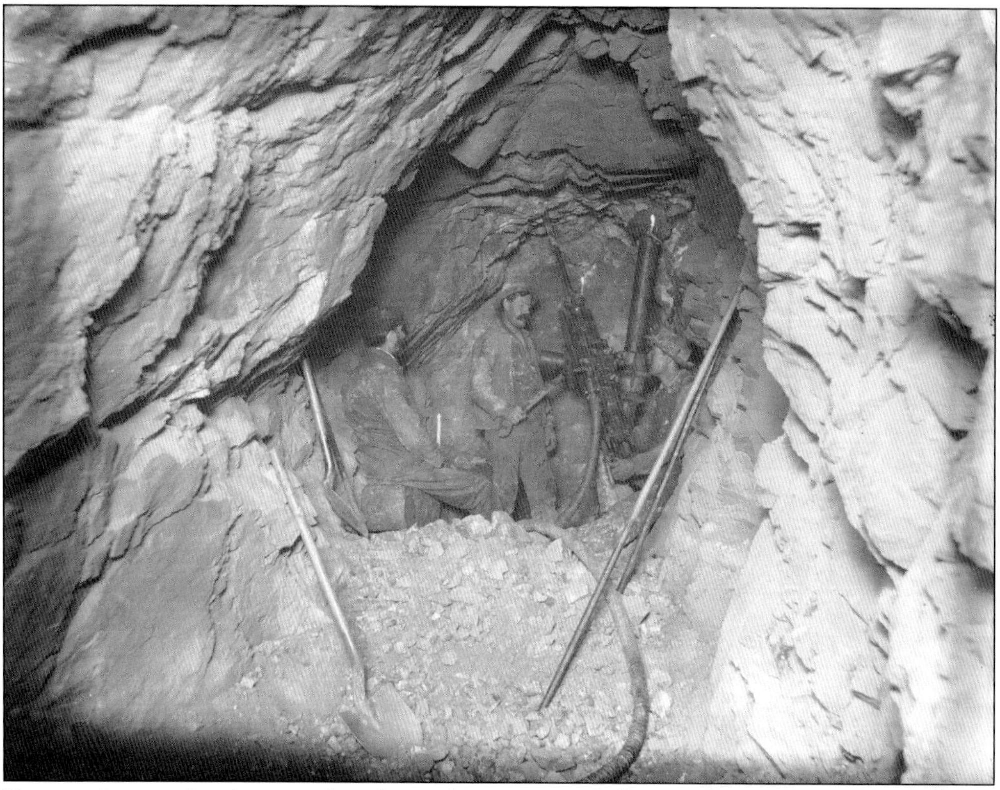

Here, an image of underground workings shows miners drilling with a pneumatic air drill. Air drills replaced the old manual single- and double-jacking methods. Compressed air was much more efficient and powerful. This meant more tunneling progress for the mining companies. Industrial progress brought many improved and necessary changes in the mining industry to make the mines and mining lower grade ore profitable. (UCB.)

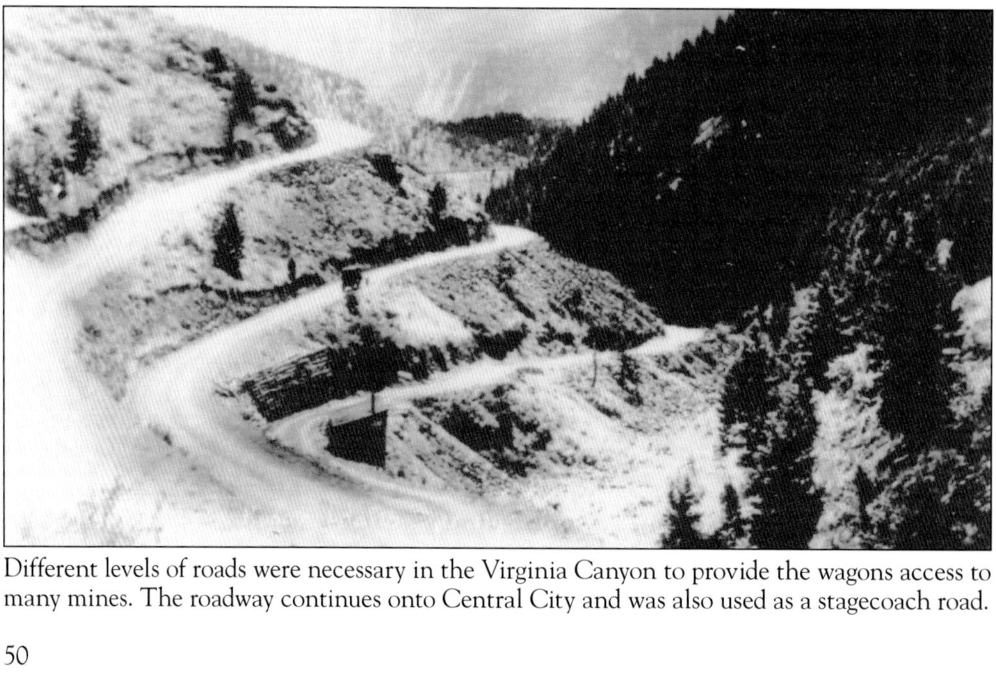

Different levels of roads were necessary in the Virginia Canyon to provide the wagons access to many mines. The roadway continues onto Central City and was also used as a stagecoach road.

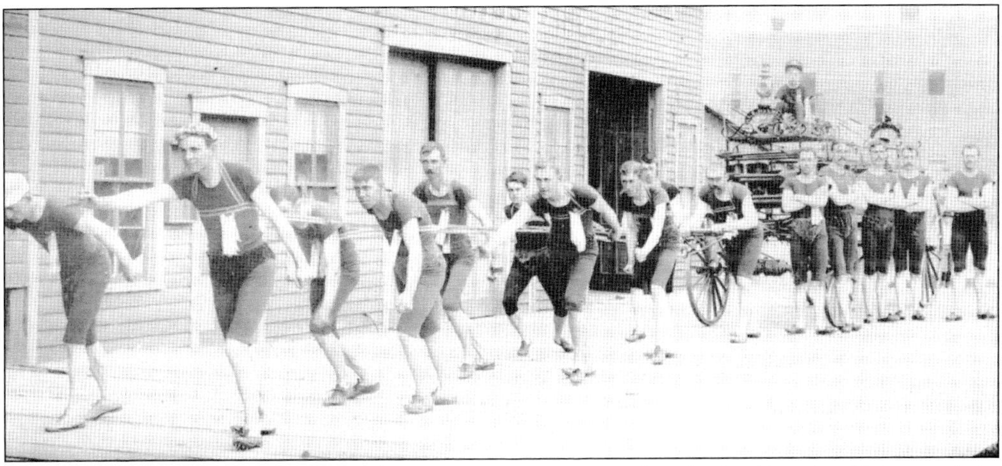

The Idaho Springs Fire Department hose cart team is strapped in for the Fourth of July parade in 1895. Two members and a small child have joined the group. Note the spiked shoes for traction to pull the heavy hand-drawn fire wagon. (DPL.)

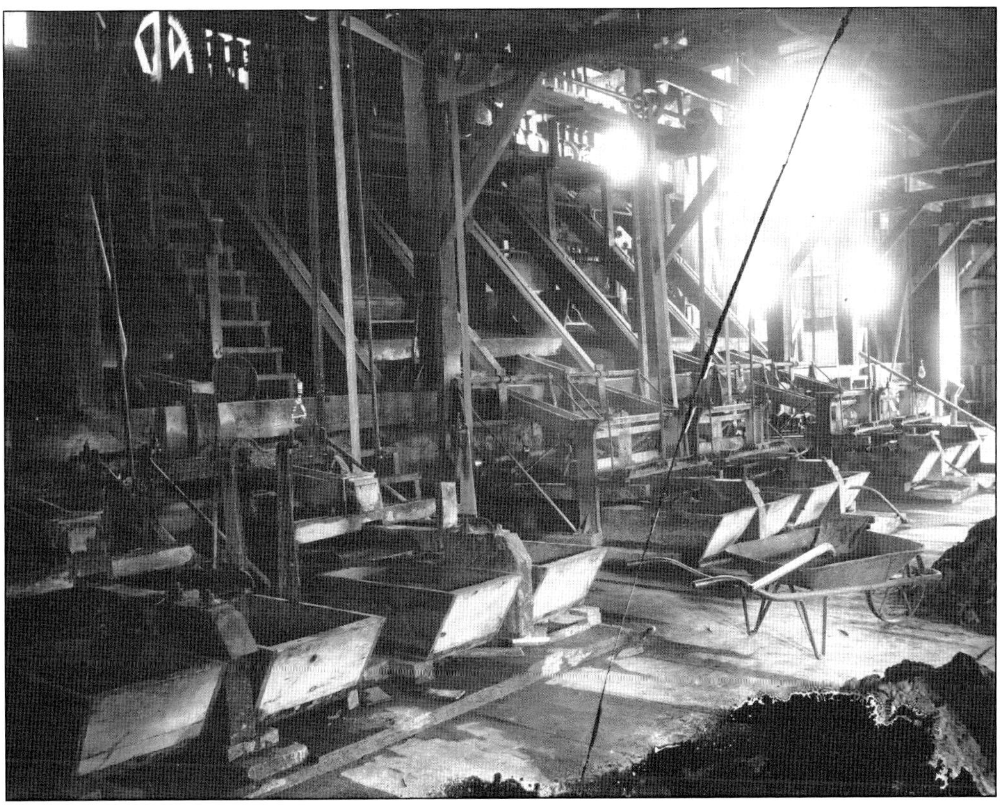

Here is an interior view of the Argo Mill. Much of mine work and mill work consisted of manual labor. The ores were moved by hand in a wheelbarrow or shoveled into an ore wagon. Here, there are piles of processed ore concentrates just run through the stamps. The next part of the mill-run process involves more crushing and a mixing (amalgamation) with chemicals or water to get the ore into a powder form so it can be shipped to the smelters for further refining. (UCB.)

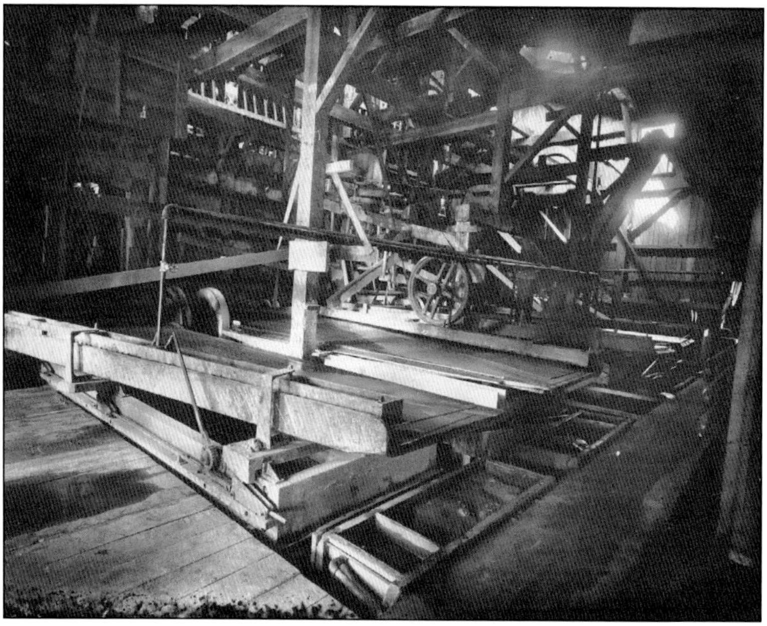

Here is an interior view of the Argo Mill and Wilfley tables. These tables would have chemicals put on them, and the processed ore would run over them. The precious metals (gold, silver, lead, and copper) would stick to the chemicals. Different chemicals were used to capture different minerals. (UCB.)

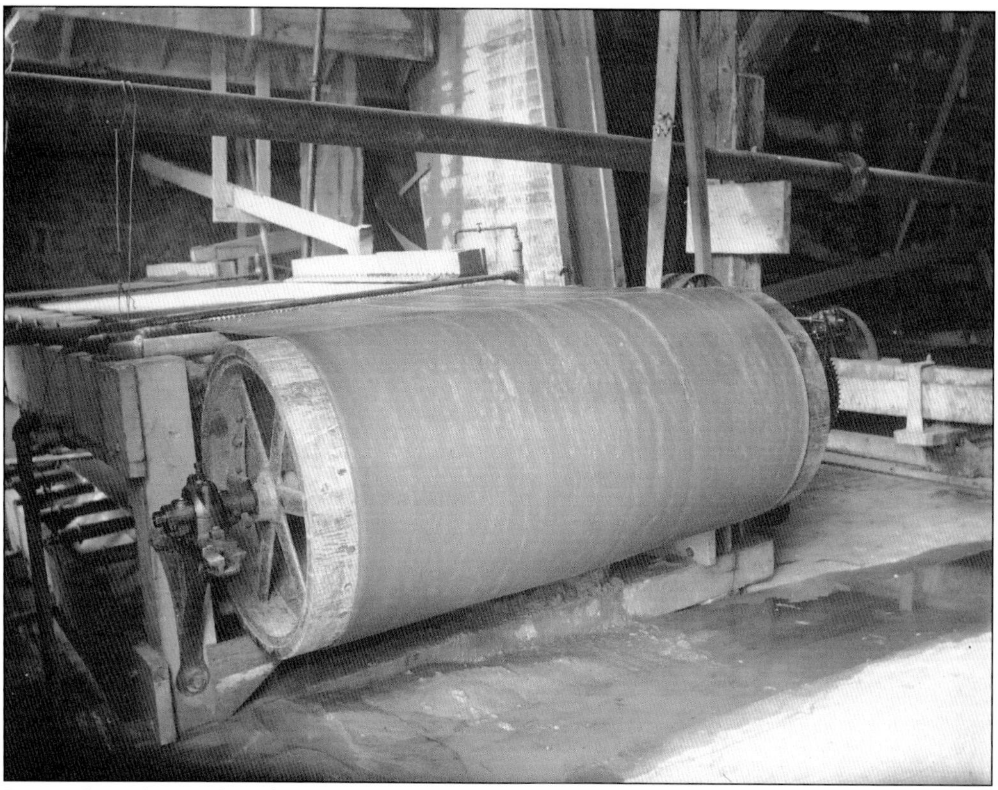

Pictured are the crushers (Frue vanners) at the Argo Mill. This improved technology increased a mill-run production. Crushers and stamp mills at most mills were often running 24 hours a day in any mining town, making the noise a part of life. (UCB.)

Three

Georgetown, Silver Plume, and Waldorf

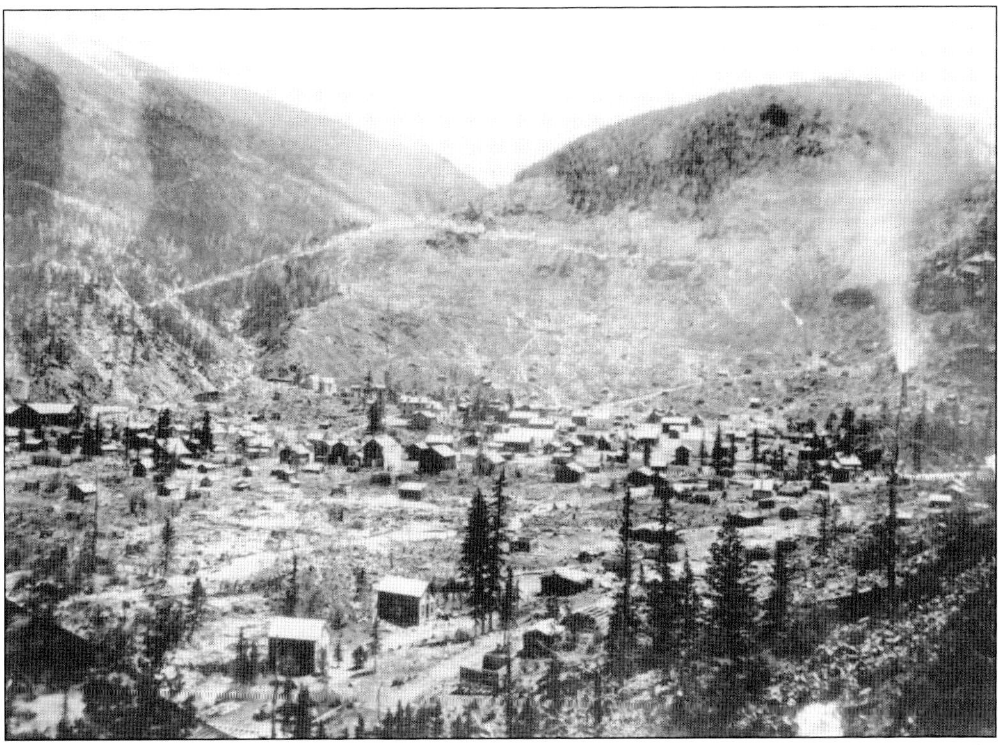

This view of Georgetown in 1867 shows a town in its infancy. The Stewart Reduction works can be seen chugging smoke as it processes the gold and silver ore that was mined nearby. Silver discoveries above Georgetown sparked a silver rush in the area. The What Cheer Mill, located center left, also processed ore from mines. "What cheer" was often used as the friendliest of greetings by Cornish miners who came to the United States following the collapse of the copper industry in Cornwall, England, in the early 1860s. They migrated to the copper mines in Michigan and all over America in search of a better life. The newly found mines of Colorado offered a new opportunity for them as well. (DPL.)

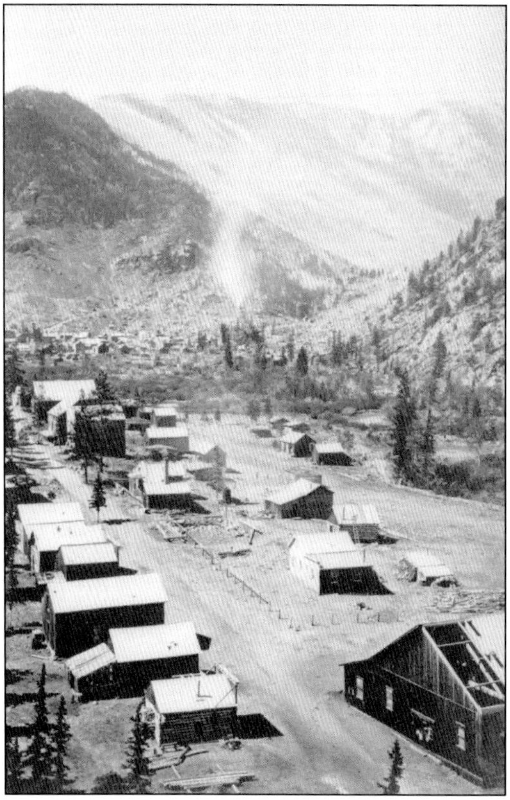

A view of Georgetown in 1867 shows growth in the Clear Creek Valley and construction of a livery barn getting a roof as the building in lower center gets some new timbering as well. The Stewart Reduction works is notable in mid-center. The mountainsides were denuded and the timber used for early construction of houses and buildings. The lower part of the valley was originally named Elizabethtown for one of the Griffith family members. The two towns later merged and became Georgetown. (DPL.)

The Barton House was located in Georgetown. This was said to be the finest hostelry in Colorado. The Barton House was destroyed by fire in 1871 and rebuilt soon after. In 1870, Billy Barton, proprietor, was sent to Denver to provide a silver spike for the ceremonial last spike of the newly connected Kansas Pacific Railroad. The spike was switched at the last minute. But it was clear that Georgetown needed access to a railroad. The railroad finally arrived in August 1877.

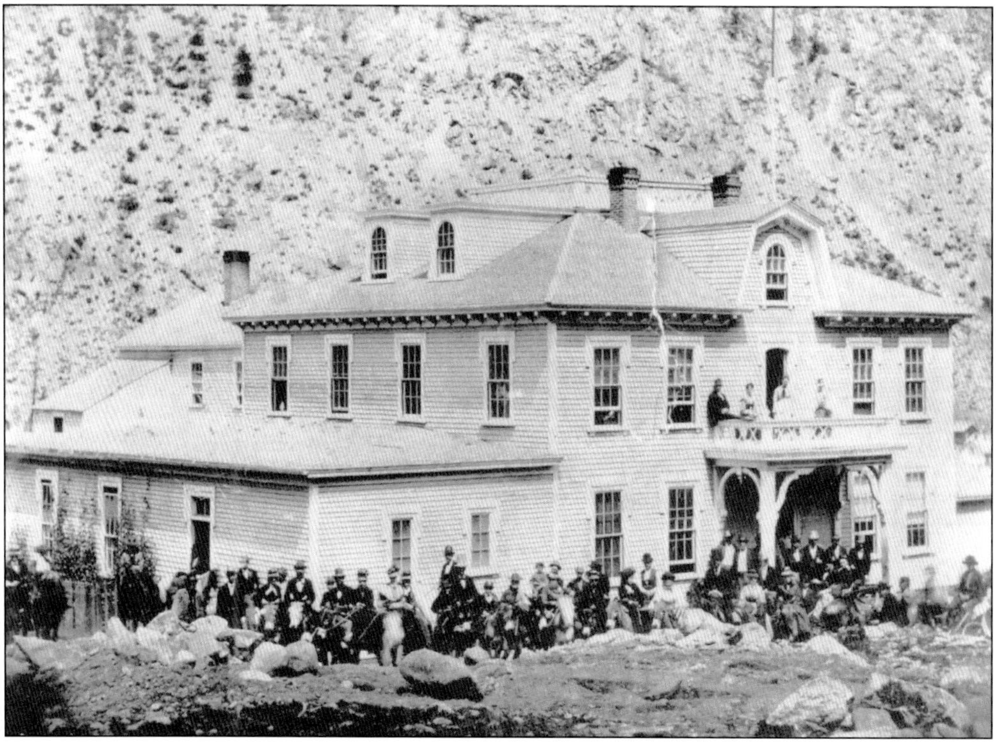

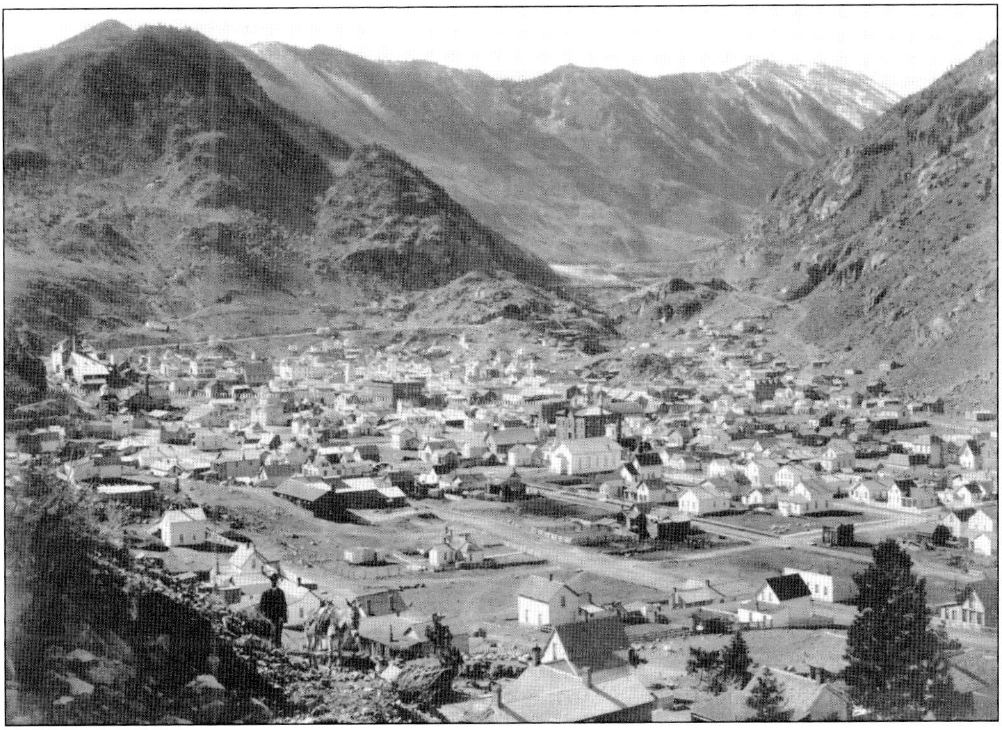

Pictured here is Georgetown around 1888. The small, rough mining town had evolved into a first-class place to live and work. The school built in 1874 is visible at right center, and the Clear Creek Mill is visible at left center. The mill processed ores from many area mines. The Hall Tunnel workings are visible above the Devils Gate Bridge, and the McClellan Opera House is still standing. This structure burned down in 1893 when Erskine McClellan attempted to thaw his interior pipes with candles. Grace Episcopal Church and the Barton House are also visible. (DPL.)

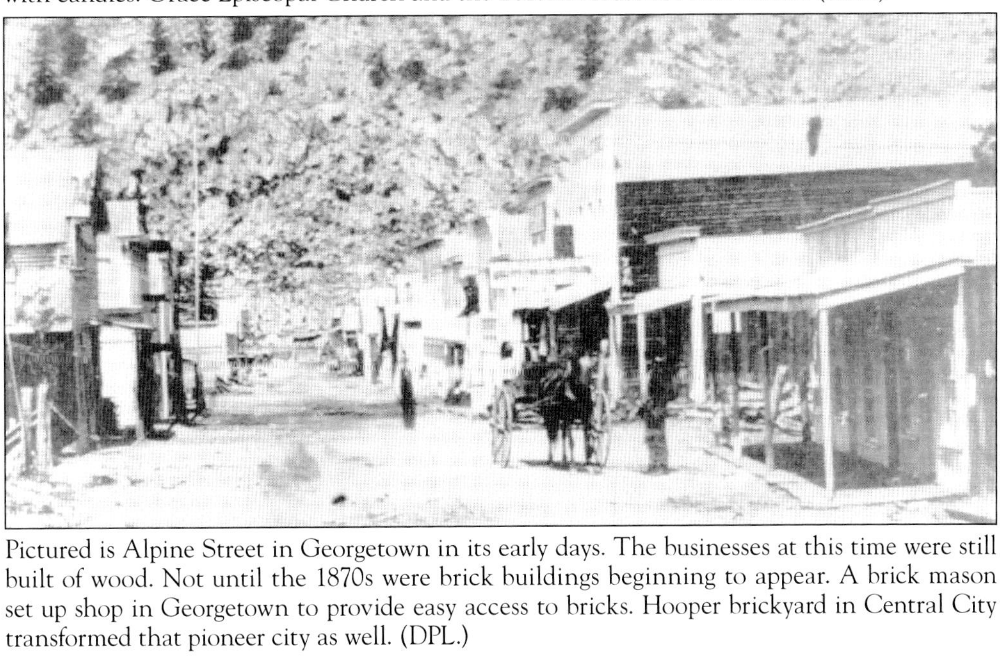

Pictured is Alpine Street in Georgetown in its early days. The businesses at this time were still built of wood. Not until the 1870s were brick buildings beginning to appear. A brick mason set up shop in Georgetown to provide easy access to bricks. Hooper brickyard in Central City transformed that pioneer city as well. (DPL.)

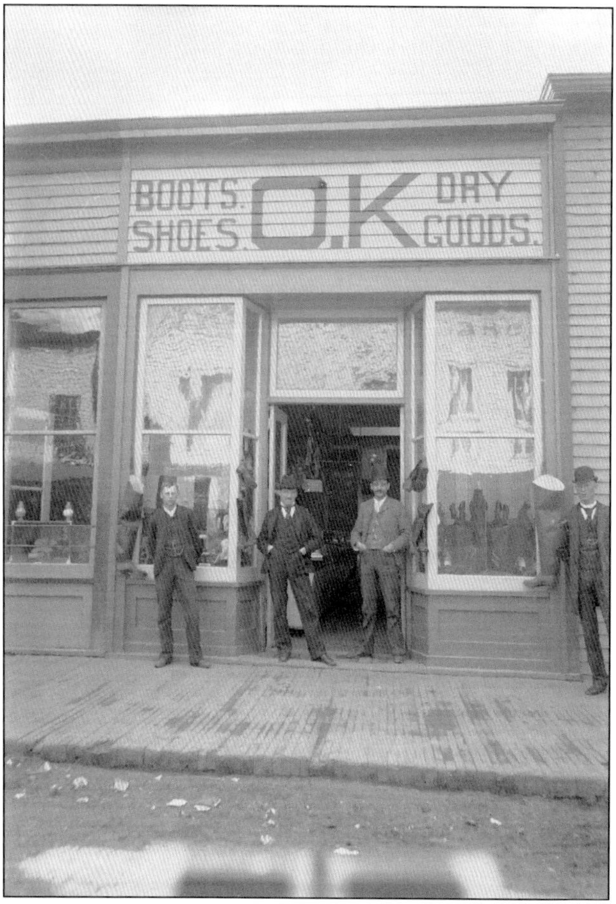

Stores such as Georgetown's O.K store were necessary to provide basic staples to their customers. The term *OK* is often said to have originated with Pres. Martin Van Buren. He was from Kinderhook, New York, sometimes referred to as Old Kinderhook. During his campaign for the 1836 election, O.K. clubs were organized. The term became to apply to something as being "all right." For example, "That guy is all right, he is Old Kinderhook." The term caught on and became popular following his presidency. There was also an OK store in Central City that sold fine clothing. (DPL.)

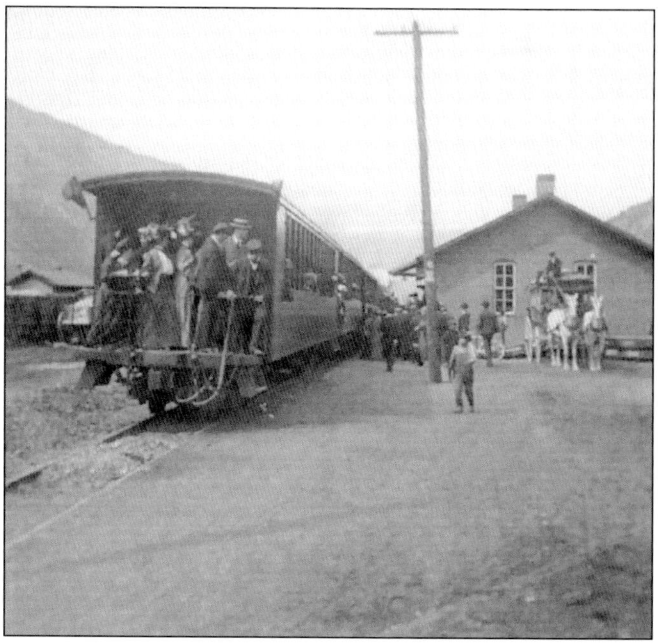

Here is a view of the Georgetown depot. Note the stage ready to depart in front of the depot. Stages would take over where the train left off. One could journey over Argentine Pass to the Snake River district or over Berthoud Pass to Middle Park. Hacking and freighting services were prevalent in the county where railroad service was unavailable. Potential investors would also be transported to the mines. Stages ran a tight schedule and would usually depart to destinations on select days of the week. (DPL.)

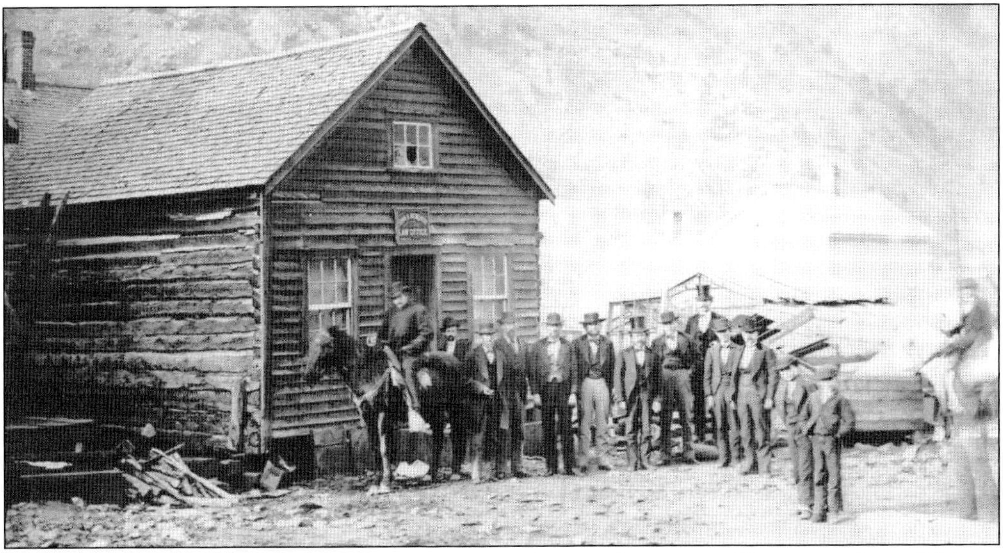

The law office of Ed Wolcott in Georgetown is seen here with a group of lawyers in front. Wolcott is third from left. He originally came to Black Hawk and worked as a teacher while his brother managed the McFarlane Foundry in Gilpin County. Ed Wolcott later relocated to Georgetown and worked as an editor for the *Colorado Miner* before becoming district attorney and later a state senator. A popular man, Wolcott eventually relocated to an area southwest of Denver where he built Wolhurst and lived there with his brother. The log cabin was located right next to the courthouse built in 1867 and was relocated next to Clear Creek when the new courthouse complex was erected in 1976. (DPL.)

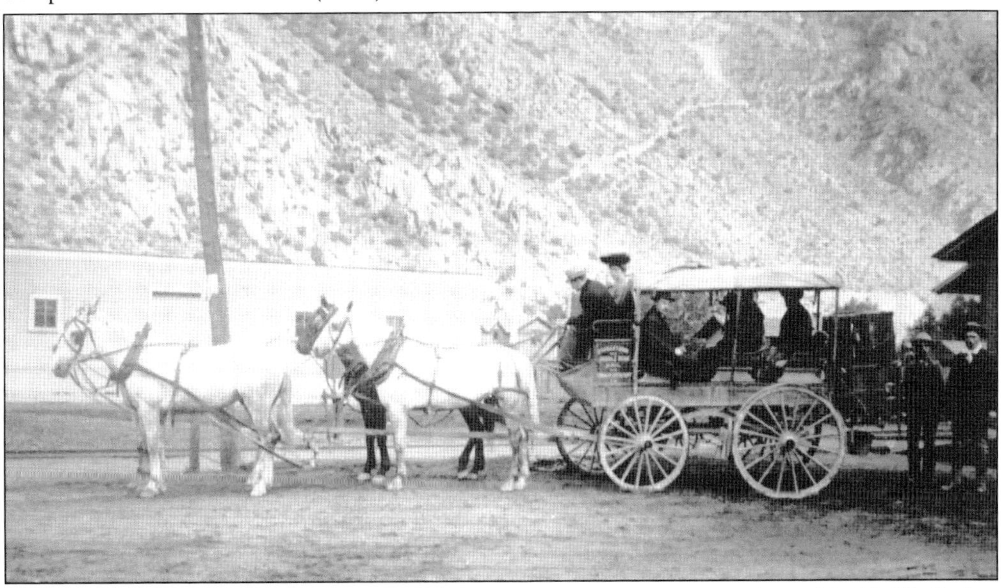

This Georgetown and Middle Park Mail and Transportation company stage bound for Middle Park is leaving the railroad station. Stage rates were not cheap, and stages generally carried wealthier patrons. Four horses provided adequate power to move the stage over the steep surrounding mountains. Georgetown bankers, to ensure Georgetown was not cut off from the transportation lanes, built the Berthoud Pass wagon road in 1874. Note the large warehouse next to the rail line. Goods were often delivered and stored in warehouses close to a town's industrial center. (CCHS.)

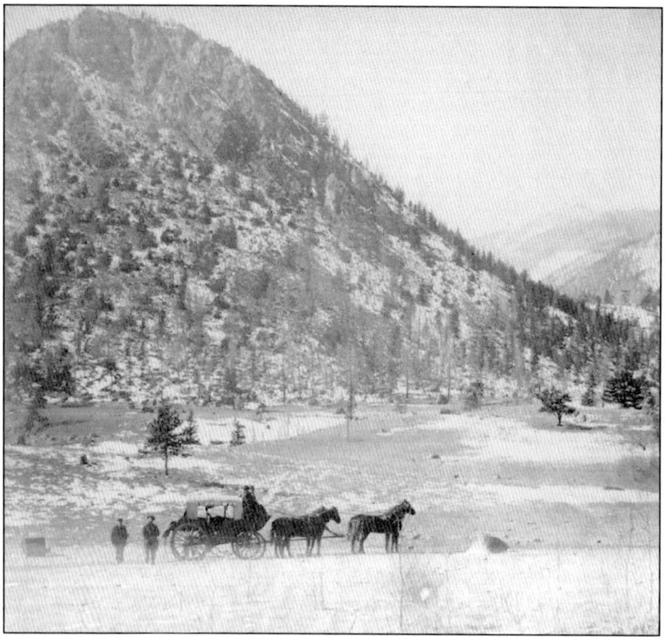

Driver Hy Washburn (behind the reins) poses with his stage leaving Georgetown at the base of Douglas Mountain below Empire. Washburn, on the Georgetown Run, and Billy Updike, on the Idaho Springs run, were masters of their trade. Speed and efficiency defined both men. Washburn was later permanently injured in a stage accident. Sen. Ed Wolcott set up a lifelong pension for Washburn. This is on the road to Georgetown in the early days before the railroad entered the valley. The interstate runs over this route now. (DPL.)

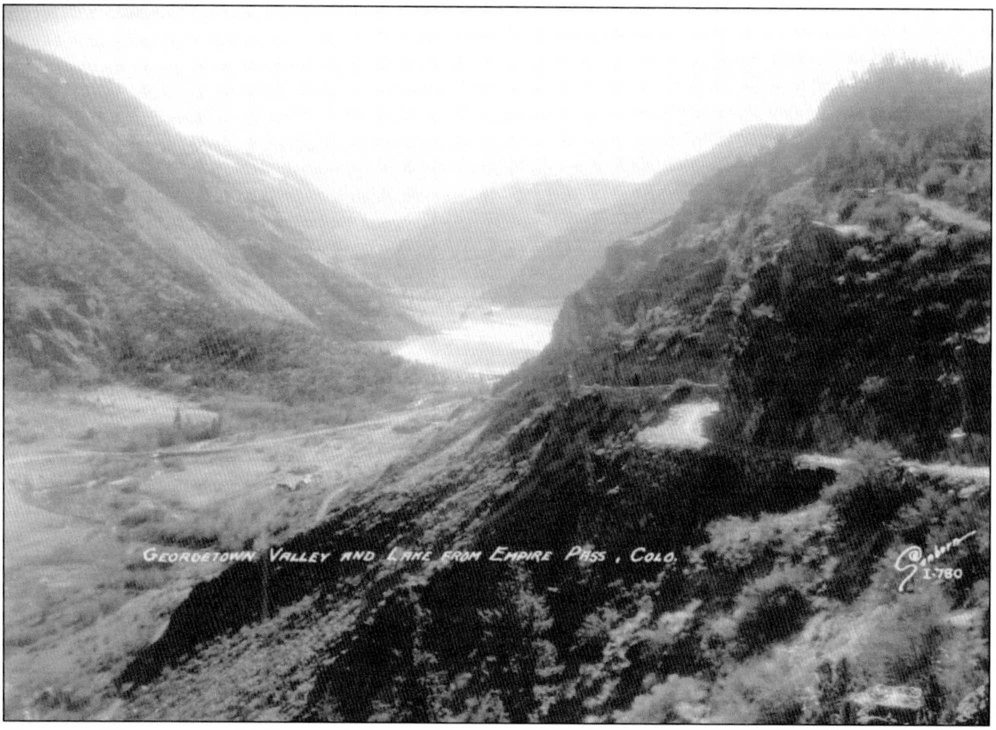

Here are Georgetown Valley and Empire Pass in the 1930s. This was the wagon road built in 1874 for the Berthoud Pass route to Middle Park. The old Silver Creek wagon road is visible at left center, and Georgetown Lake in the middle covered the old mill tailings. The dam broke in 1956 and again in 1957. The road to the upper right provided access for the mines up Bard Creek. Camp Beshears and other areas saw high productivity in the early 1900s. Old mine and mill sites are prevalent in the area among the trees. (CCHS.)

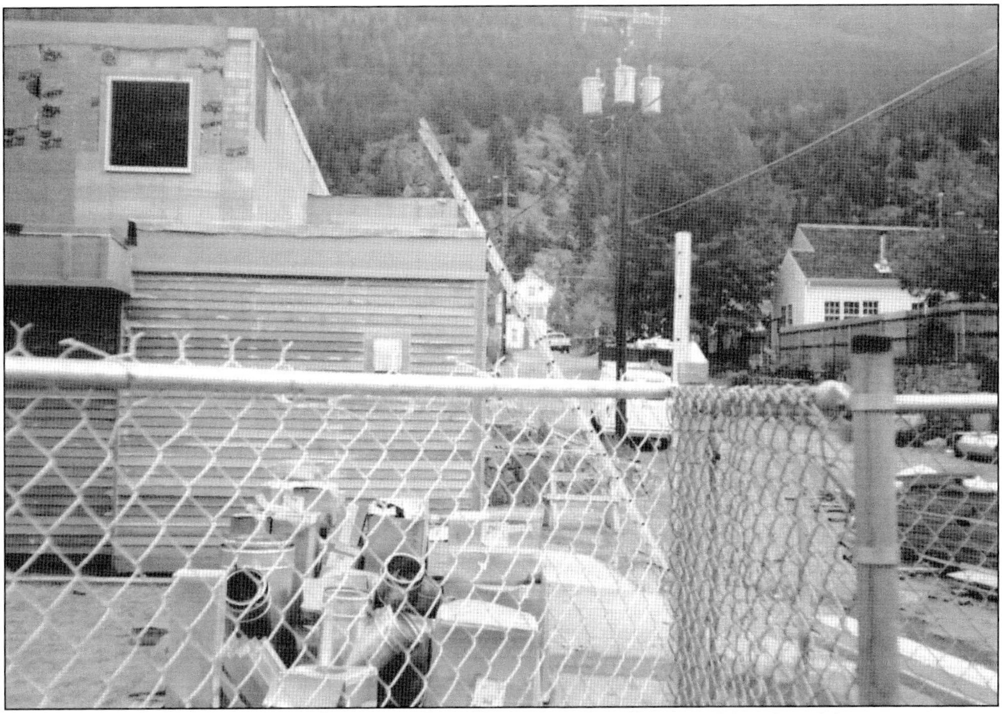

In 1996, the old school, built in 1938, was covered to make it look like the old Barton House, which had originally stood at that location. The school was erected in 1938 as a Works Progress Administration (WPA) project at the same time the brick school building was constructed in Empire, replacing its 1897 school. The original brick Georgetown school, built in 1874, was condemned. The new school served as the high school and became Georgetown Elementary after the Clear Creek School District reorganized, making elementary schools out of the schools in Georgetown, Empire, and Idaho Springs. Georgetown Elementary is where the Georgetown Society undertook the Georgetown Guides project in the late 1970s. (CCHS.)

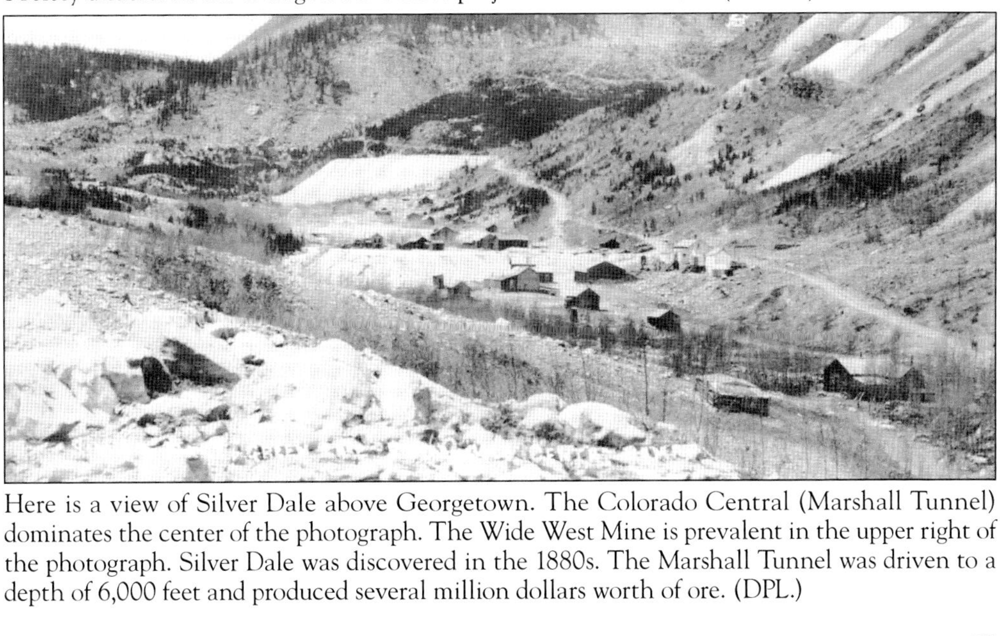

Here is a view of Silver Dale above Georgetown. The Colorado Central (Marshall Tunnel) dominates the center of the photograph. The Wide West Mine is prevalent in the upper right of the photograph. Silver Dale was discovered in the 1880s. The Marshall Tunnel was driven to a depth of 6,000 feet and produced several million dollars worth of ore. (DPL.)

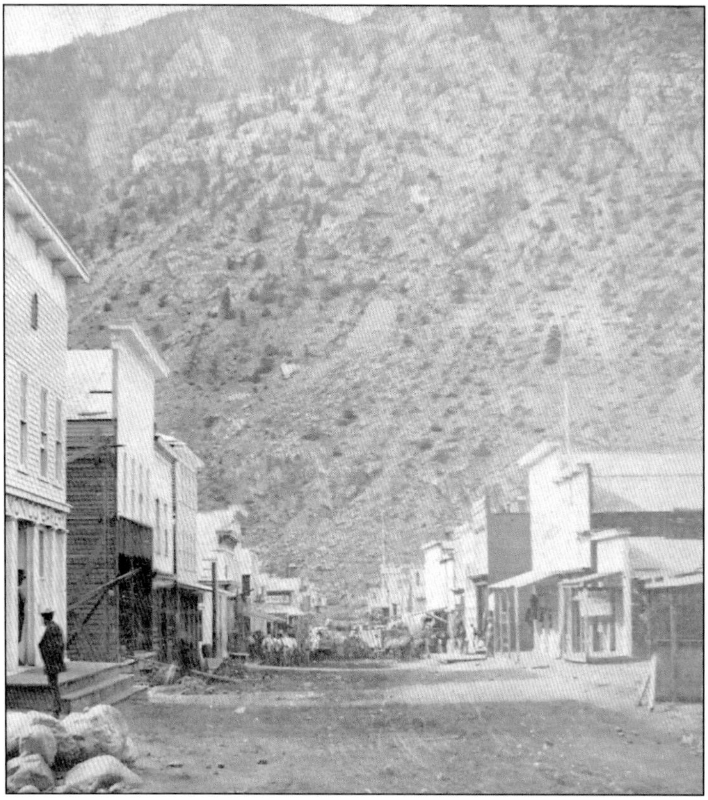

Here is Alpine Street in Georgetown's business district. The Delmonico Bakery on the left was purchased and rebuilt into the Hotel De Paris by Louis Dupuy in 1875, following a mining accident in which Dupuy was severely injured. There are ore wagons on the street, and the McClellan Opera House (fourth building on the left) served as a point of entertainment for Georgetown residents for many years. (CCHS.)

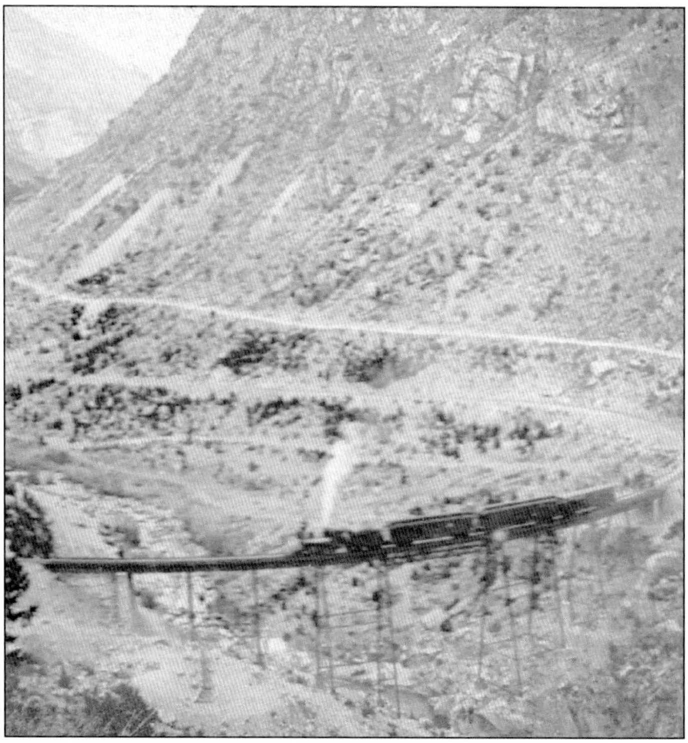

The Georgetown Loop and Devils Gate Bridge are seen here in 1885. Jay Gould and the Union Pacific Railroad built the Devils Gate Bridge when the push was on to get to Leadville and service the silver rush going on there. The original bridge was installed incorrectly in 1884 and had to be refitted and placed. The Georgetown Loop served as a way to access the mines of Silver Plume and became a major tourist destination in Colorado. (History Colorado.)

The Georgetown Loop is a narrow-gauge railroad and currently operates as a tourist railroad. It was engineered by Ed Berthoud and accessed the thriving mining towns of Georgetown and Silver Plume. A corkscrew was designed to elevate the railroad to 600 feet. The Georgetown, Breckenridge & Leadville Railroad had been formed in 1881 under the Union Pacific Railroad. There were plans to continue the line up and through the mountains to the west. The Rio Grande Route reached Leadville first from Salida, thwarting any plans of reaching Leadville via Georgetown. (History Colorado.)

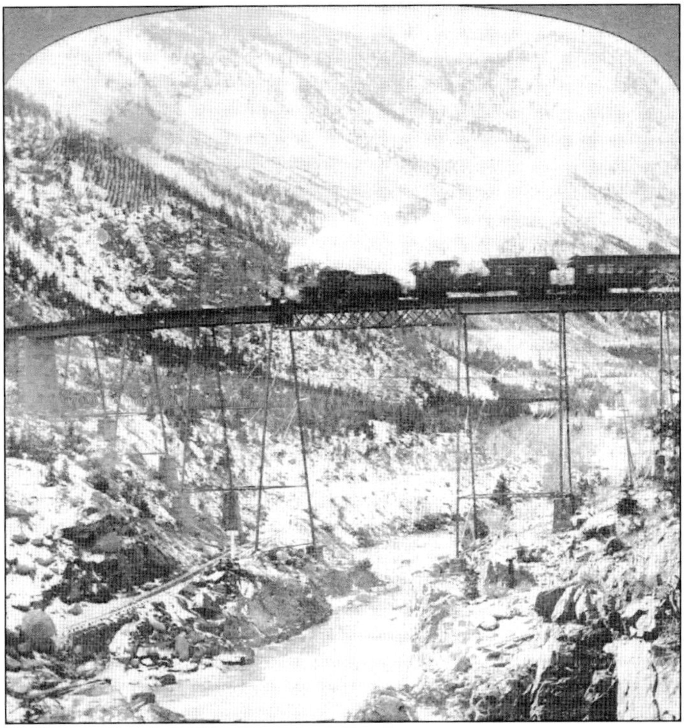

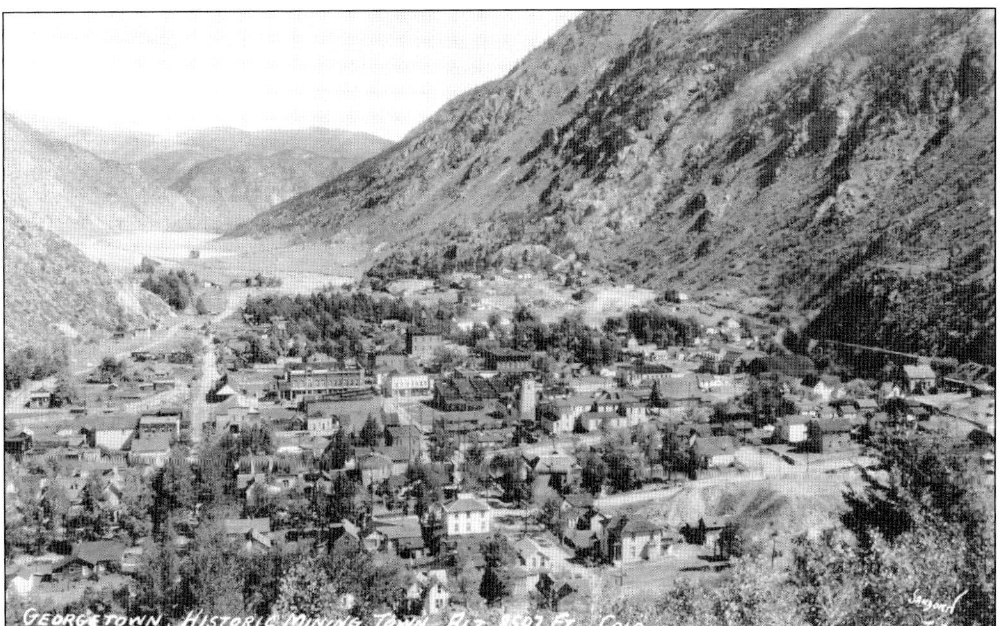

Georgetown is pictured here in the 1920s. Note the Barton House right center next to the Alpine Hose Tower. During these years, it was virtually abandoned and left to decay. It was razed in the 1930s, and a new school was built at that site. Note the railroad grade to the right. (CCHS.)

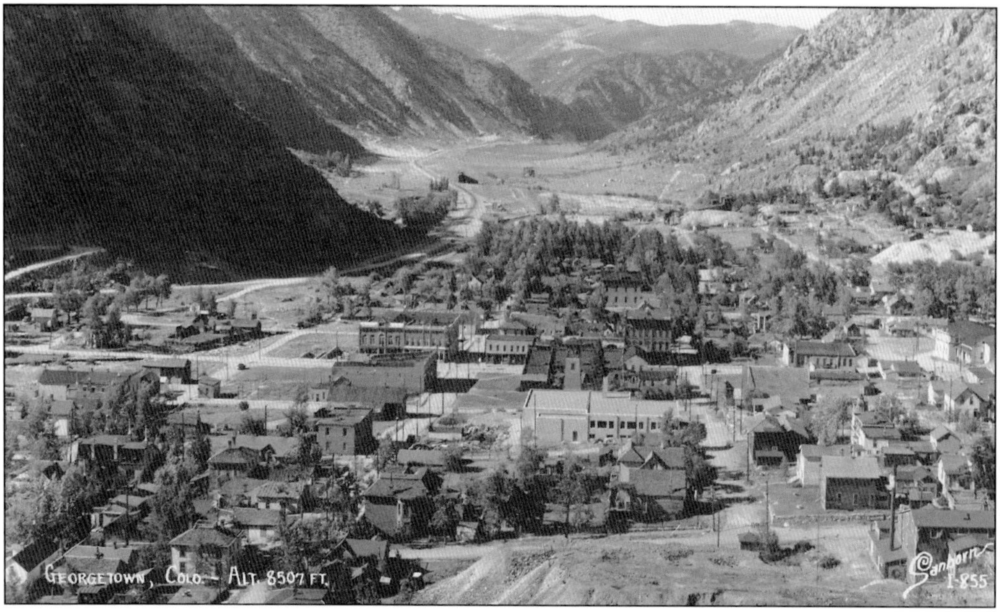

This Georgetown view was taken after 1938, following the construction of the new school on the old Barton House site. Note the construction of the Centennial Mill across from the house with three windows. This was built when the price of gold went from $20 to $35 an ounce in 1934 and enabled older nonproducing mines to reopen. Compare this area with the same area in the previous photograph. Note that the tower on top of the old 1874 school has disappeared. (CCHS.)

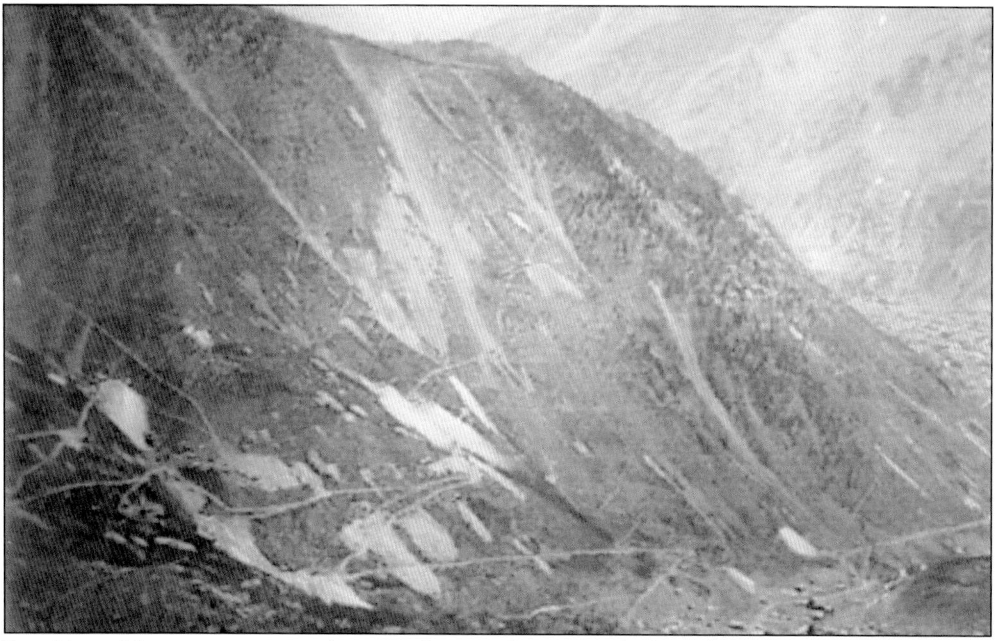
Republican Mountain is seen in this view showing the mines of Silver Plume and looking down from Sunrise Peak in 1908. Georgetown is at right center. The Georgetown Loop ran from Georgetown to a higher elevation at Silver Plume. The tracks loop over themselves to gain elevation. Note the zigzagging wagon road originating from Silver Plume to access the ores of the mines. These roads would be very dangerous during the winter months.

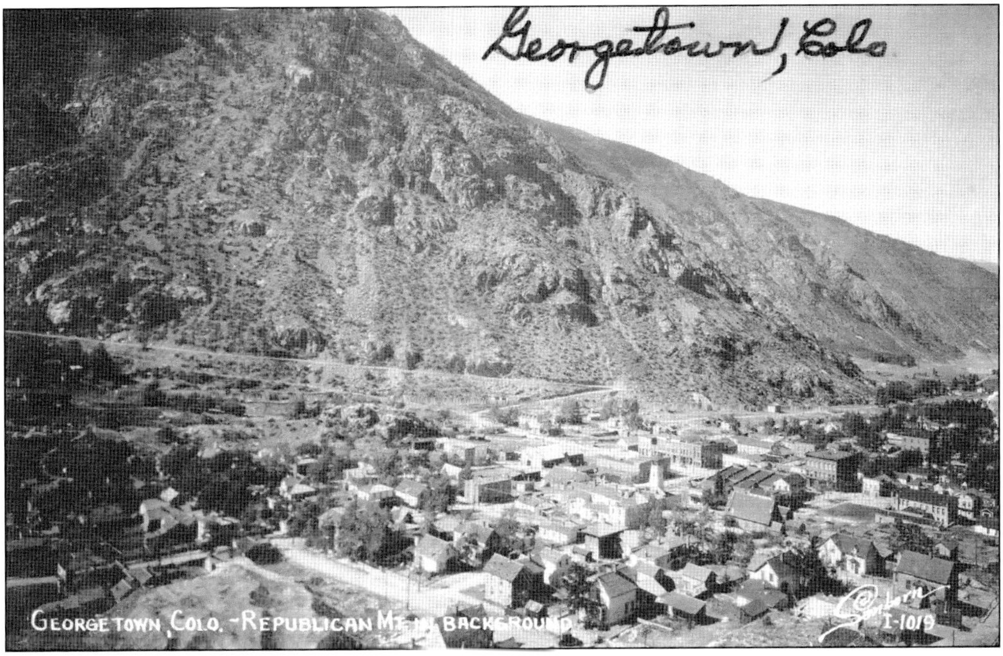

These 1930s views of Georgetown show the Barton House is still next to the Alpine Hose, and the old courthouse is still in its original location (photograph center). The interstate would later be built where the zigzagging road to Silver Plume is. Also note the existence of the tower on the old school and the three stories on the Silver Queen Building. (Both, CCHS.)

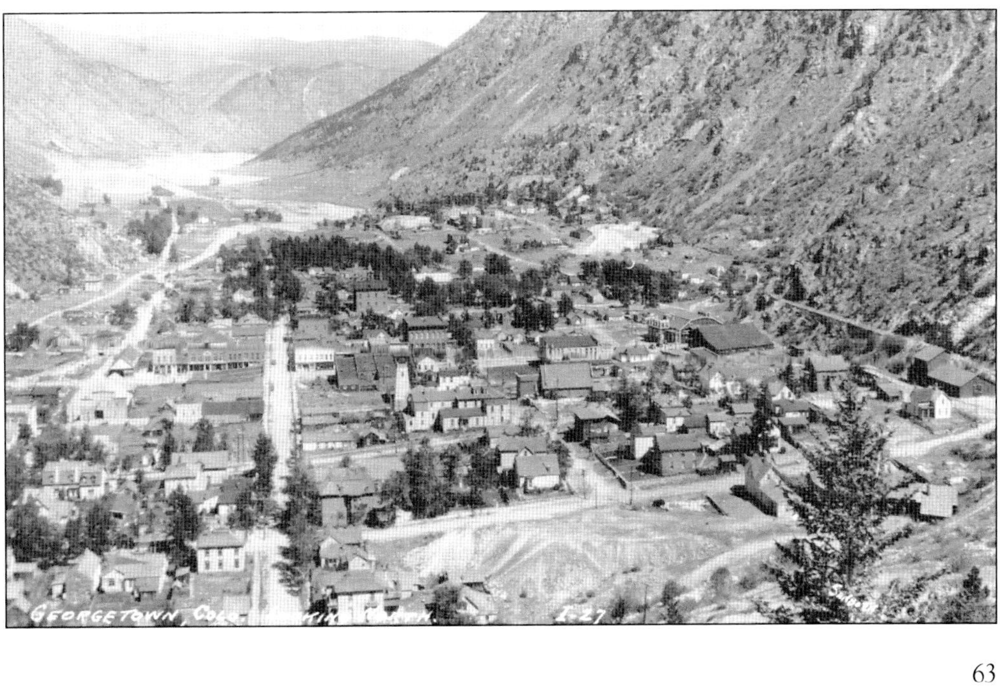

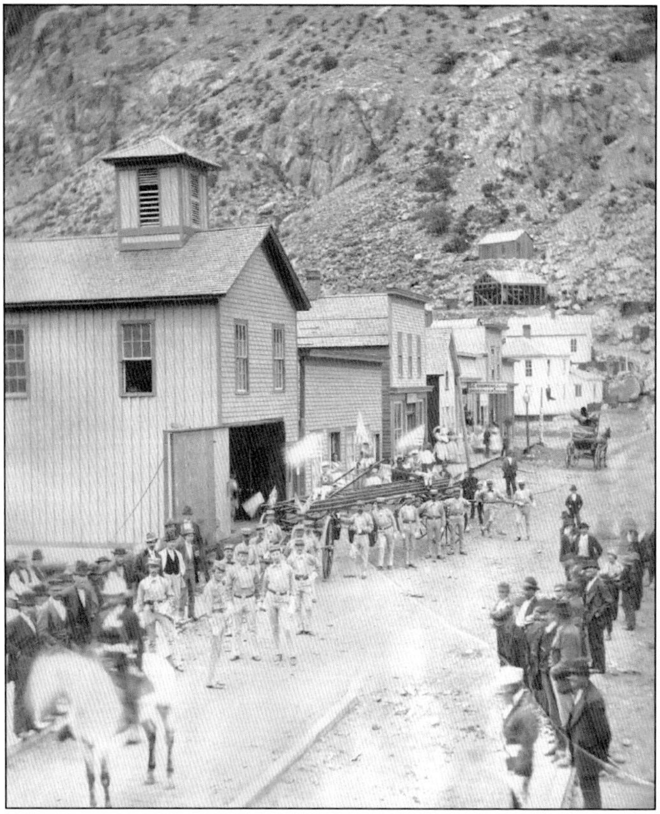

Hope Hose fire team is lined up on Alpine Street in Georgetown. Hope Hose was one of the four Georgetown fire departments that kept the town from suffering a major conflagration during the course of its existence, beginning in the 1860s. By the 1870s, fire control became a major concern for any pioneer town. Pioneer mining town Central City also suffered a major fire in 1874. It was rebuilt with brick, and the town continued to prosper after suffering the great calamity. (DPL.)

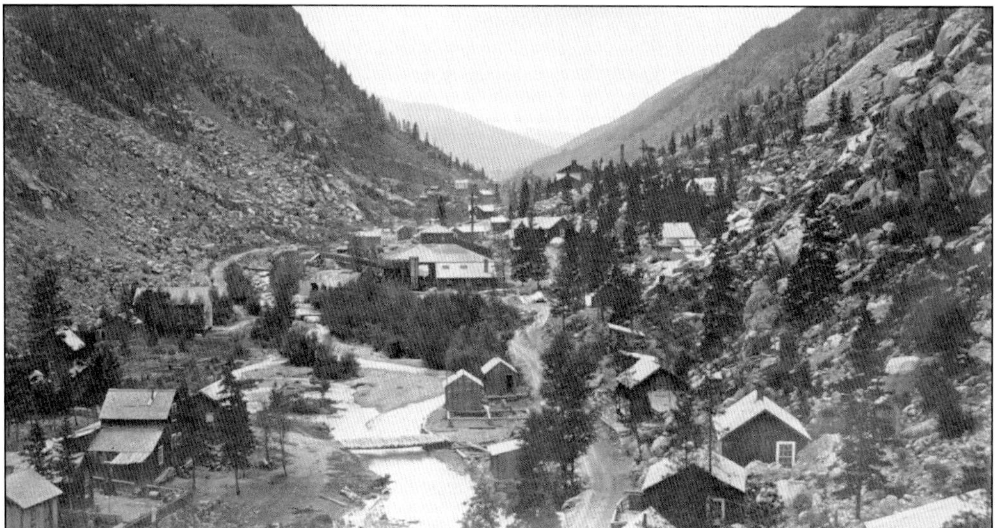

Brownsville is seen here in 1870. This town was located up from Silver Plume and was completely covered by a mudslide from the Seven-thirty Mine, located up above in Brown Gulch (to the right). The Terrible Mine was developed atop the slide and can still be seen today. Interstate 70 made a path right through the mudslide when it was constructed through the area in the 1960s. (DPL.)

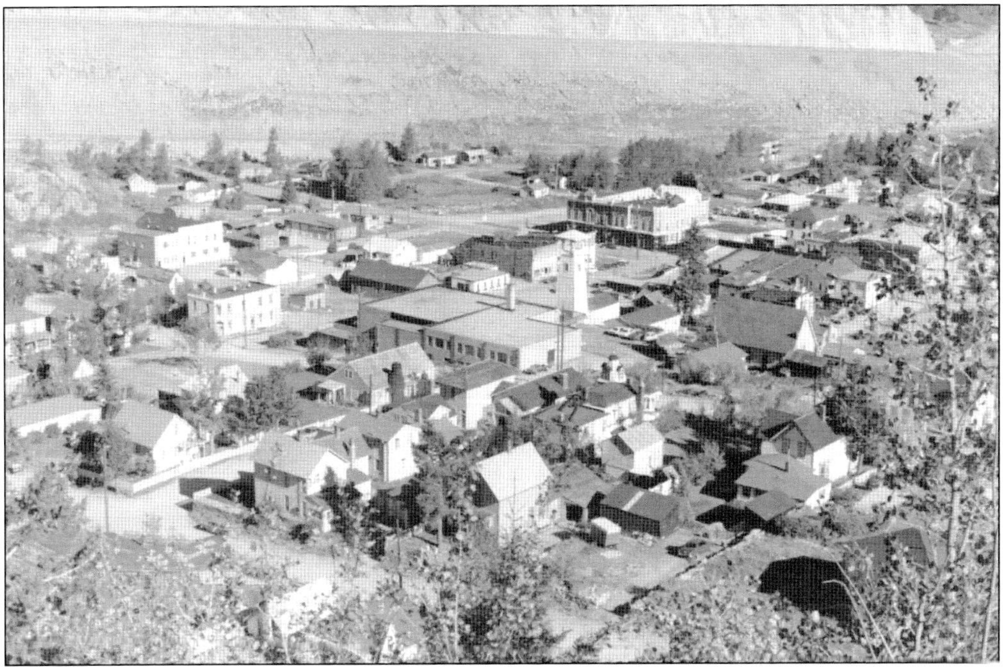

This 1975 view of Georgetown is the same view as Sanborn 1019 (page 63, top), showing the new 1938 school (center), only two stories on the Silver Queen Building (right center), and the courthouse complex (left of center). The new Interstate 70 can be seen in the background, and the police station is the small building in upper left of center. This is where Deputy Sheriff Michael Dugan worked before the new 1976 courthouse complex was erected in 1976. (DPL.)

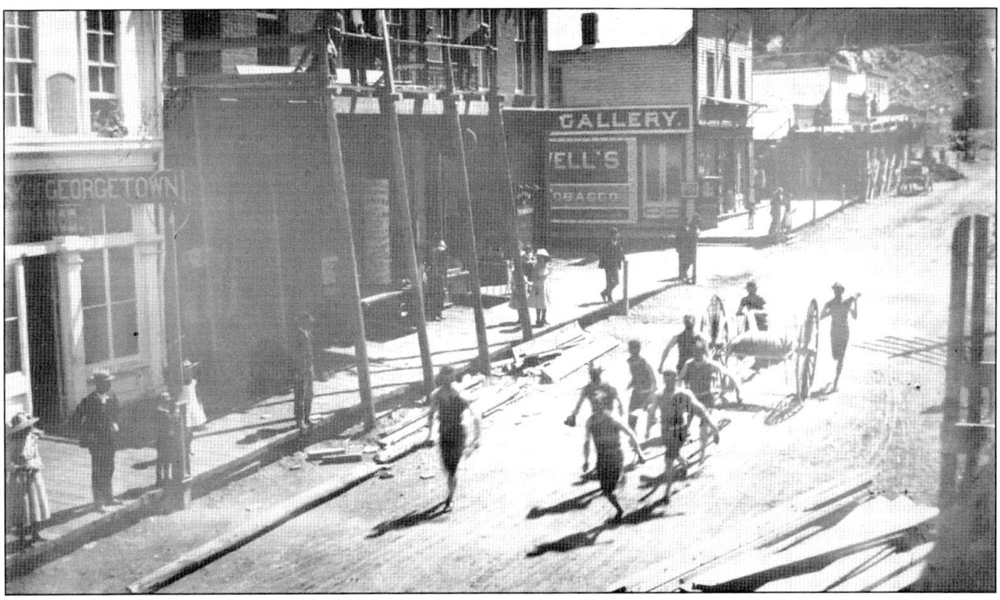

Here are hose cart races along Alpine Street in Georgetown in the 1870s. The Bank of Georgetown can be seen on the left. (DPL.)

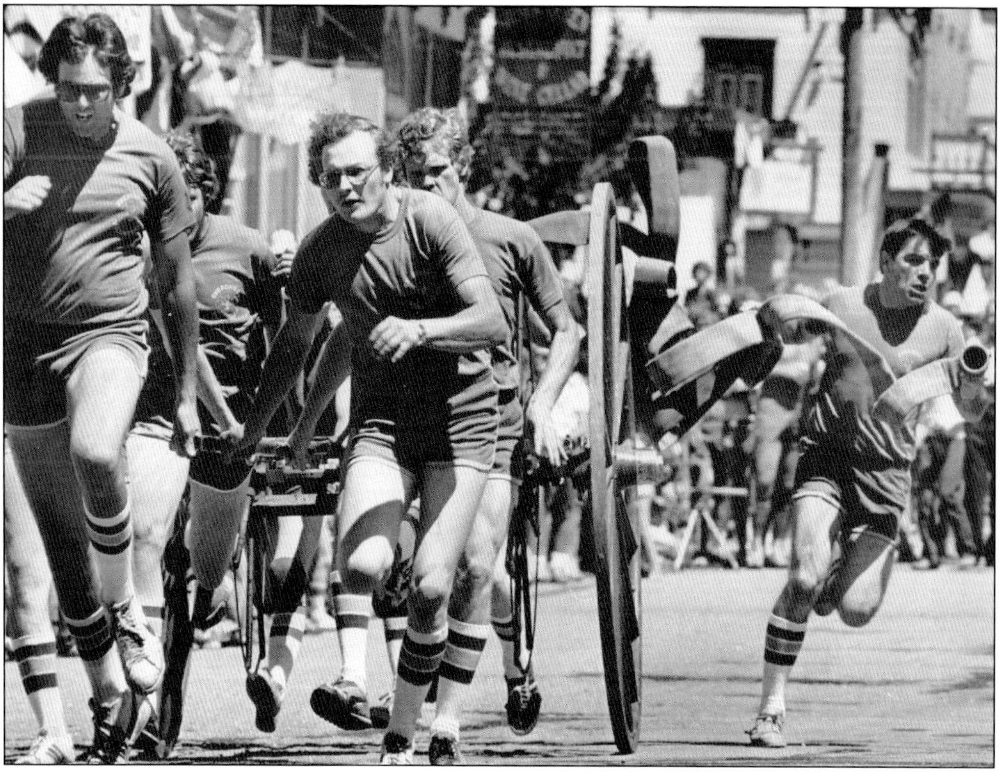

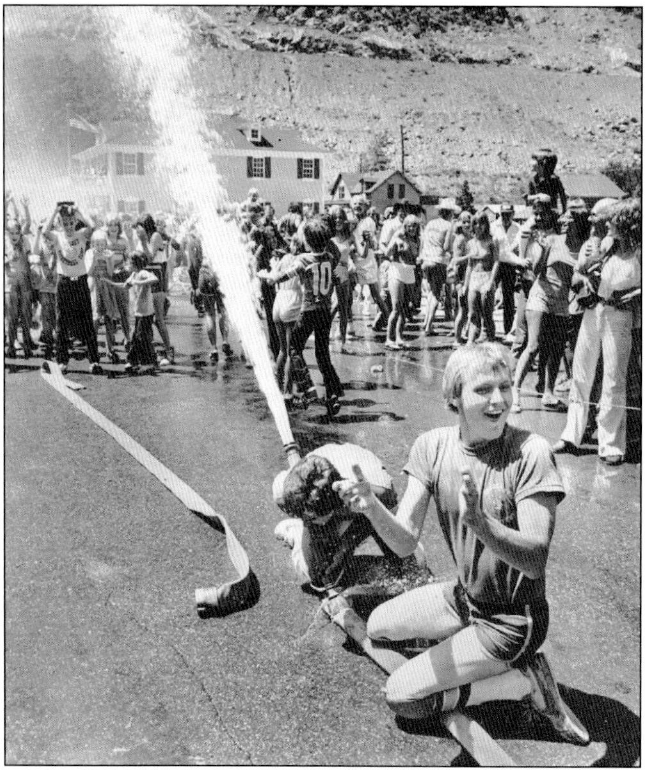

These hose cart races were held in Georgetown on July 4, 1980. The above photograph, taken almost 100 years after the one on page 65, shows the runners in virtually the same spot as the early runners. The hose cart races were the highlight of the Fourth of July in Georgetown. The end result and goal of a hose cart race is to connect to the source of water and be the fastest to shoot water out of the hose. Due to the dangerous nature of the races, they were discontinued in the 1980s. Bucket brigades have replaced the hose cart races. The post office in the background was built to emulate the old Yates House, which was located on that site in the 1870s. It is a virtual replica of the original building. (Both, *Denver Post* photographs, CCHS.)

The Georgetown Loop is seen here in 1926. The rails were taken away in 1939 following the demise of the Colorado & Southern Railway. The Colorado Historical Society formed the Historic Georgetown Loop Historic Mining & Railroad Park in 1959. The restoration of the Georgetown Loop followed, and in the 1970s, the Seabees reconstructed the loop. The new Devils Gate Bridge was built in 1984 and dedicated by Gov. Richard Lamm.

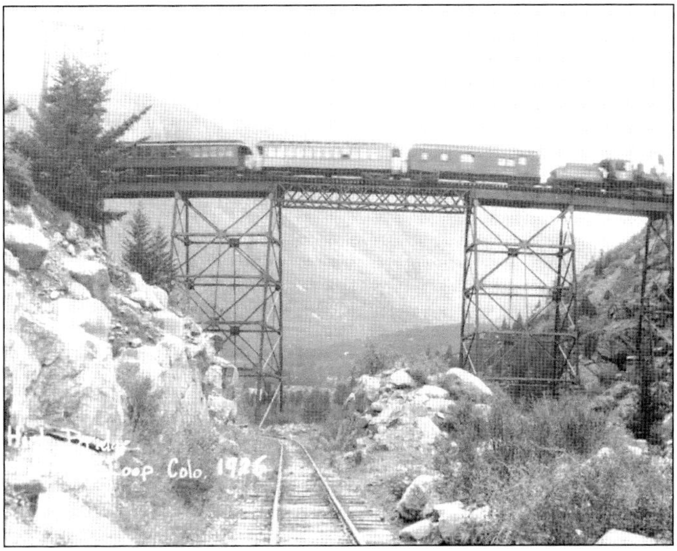

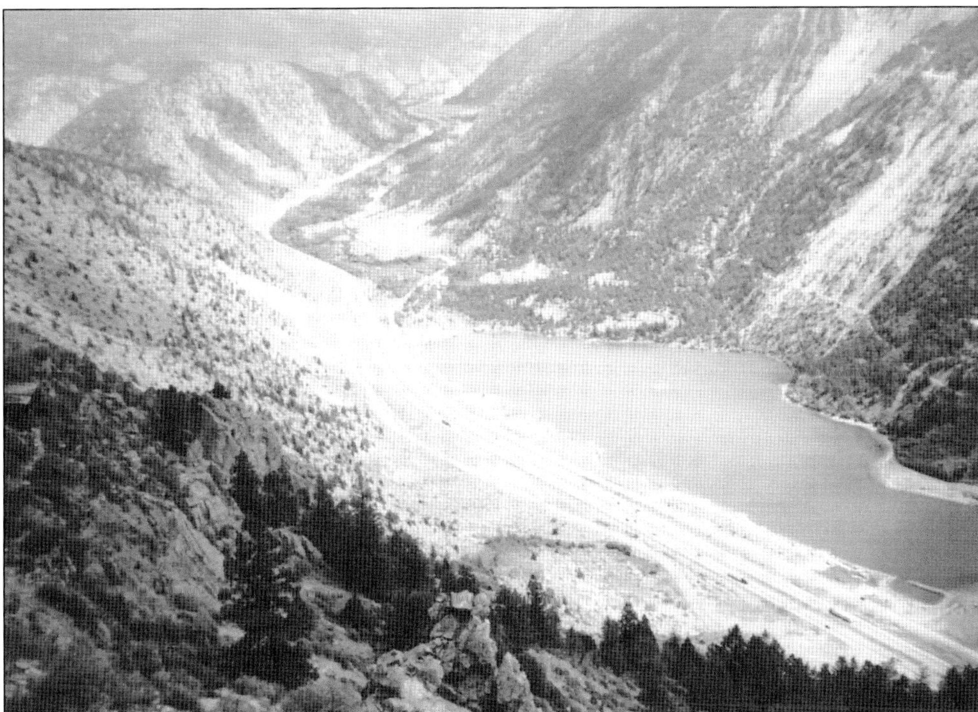

This view of the Georgetown Valley shows the Hangover Mine (left center), Georgetown Lake, and the old Silver Creek Road that was never quite completed or utilized (right center). It was built in the 1890s as a railroad grade to the silver mining town of Silver Creek but was never fully finished following the demonetization of silver in 1893. Empire and its mines are on the other side of Douglas Mountain (upper left). (CCHS.)

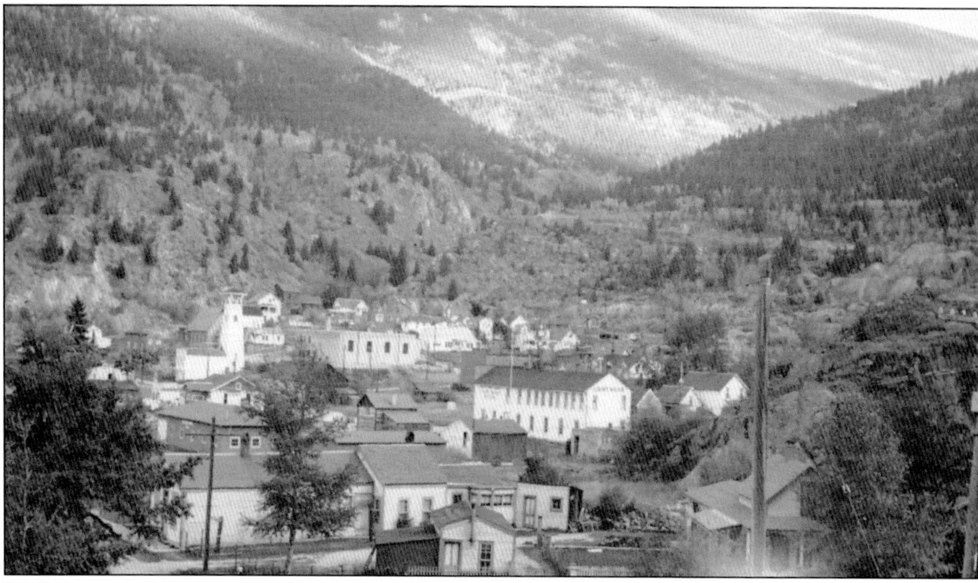

In this view of Georgetown looking southwest, note the location of the courthouse (lower center, with name on building) and the legal office of Ed Wolcott (log cabin with chimney and single window, lower center, left of courthouse) as well as the old jail directly behind (lower center right, next to courthouse). They are all within short walking distance of one another. The old cabin and old jail were relocated when the new courthouse complex was built in 1976. (DPL.)

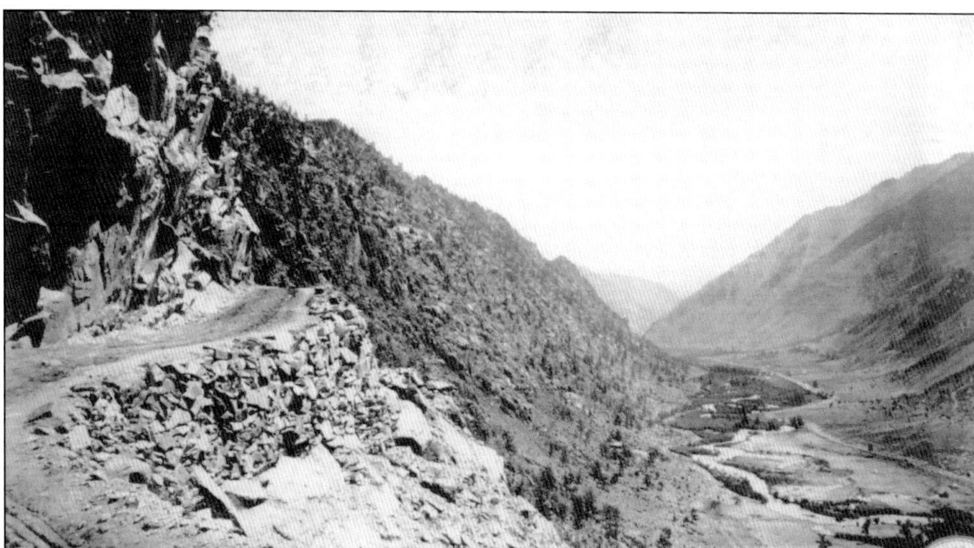

Pictured here are Empire Pass and the Clear Creek Valley. The road to Georgetown can be seen on the lower right. The view shows the valley before the Silver Creek Road was later built in the early 1890s. (USGS.)

The Alpine Hose is pictured in 1940. Georgetown had four fire companies to protect the district: the Hope Hose, Star Hook & Ladder, Alpine Hose, and the Old Missouri Fire Company. The Alpine Hose was built in the 1870s, and the tower was constructed later, after a sizable donation from William Hamill. Hamill also donated the bell in the tower. The Georgetown Courier Building is on the left, and Jesse Randall generated the newspaper there from 1877 to 1959. (DPL.)

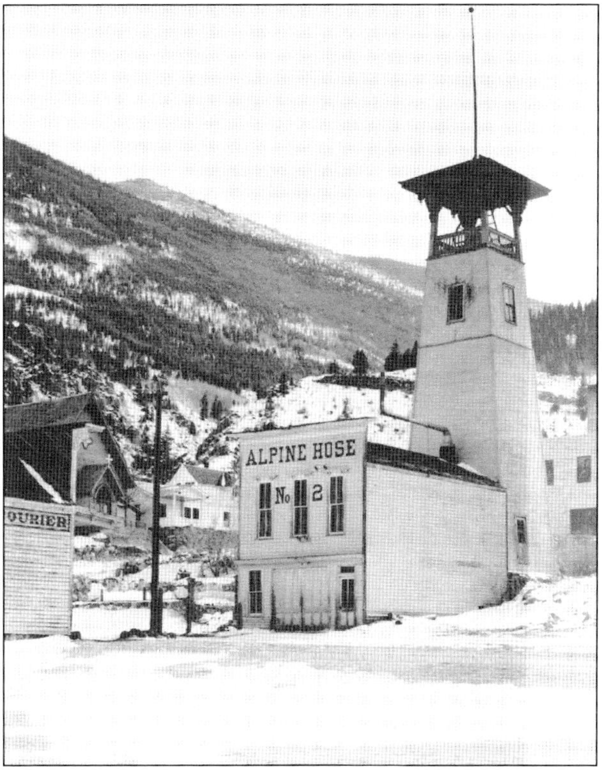

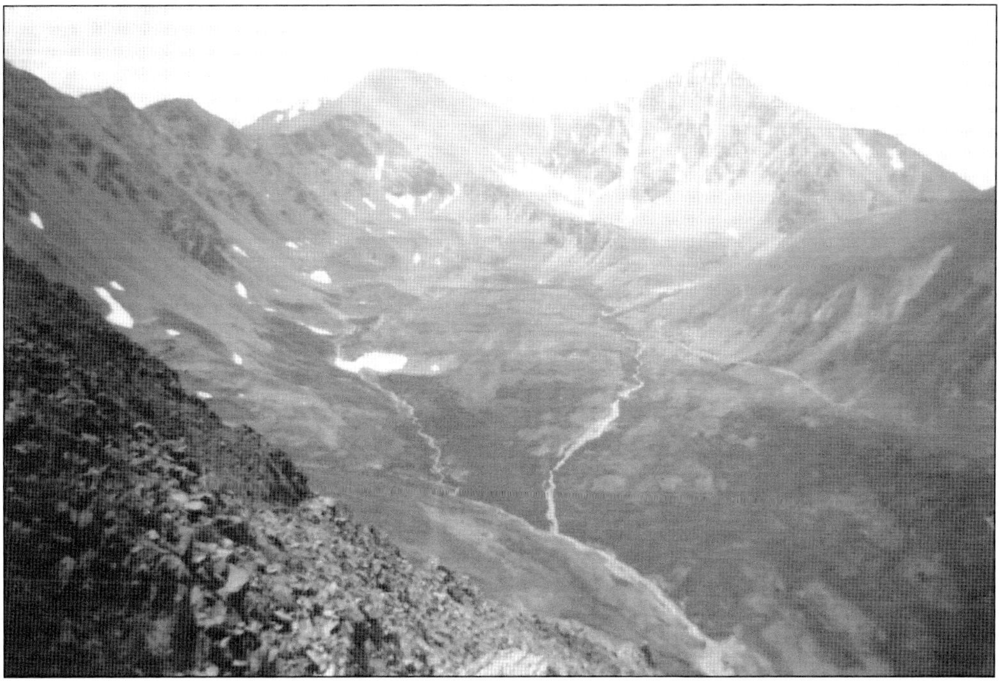

Grays and Torreys Peaks are beyond the crest from Waldorf; mining activity was prevalent in this area of the county. The Stevens Mine is below (lower right) in the valley. This is where the Argentine Central Railroad ended after climbing from Silver Plume. (DPL.)

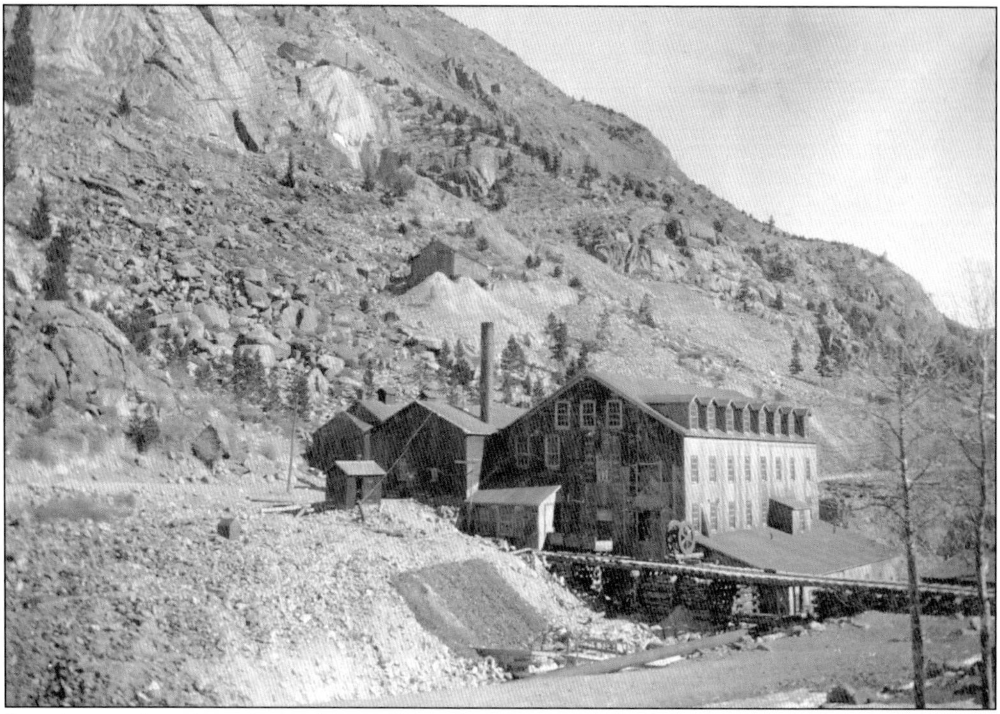

Seen here is the enlarged Mendota Mill in Silver Plume. The mill processed ore from the Mendota Mine, the Victoria Tunnel, the Terrible Mine, and the Smuggler.

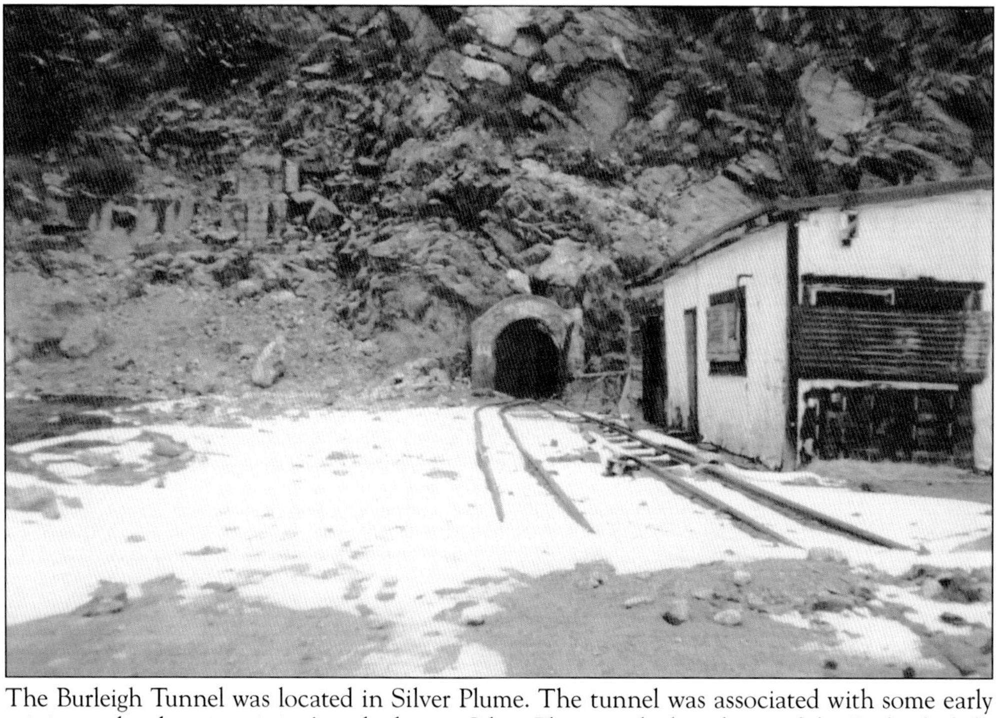

The Burleigh Tunnel was located in Silver Plume. The tunnel was associated with some early mining technology in mining's early days in Silver Plume with the advent of the Burleigh drill, invented by owner William Burleigh, who was from England. (CCHS.)

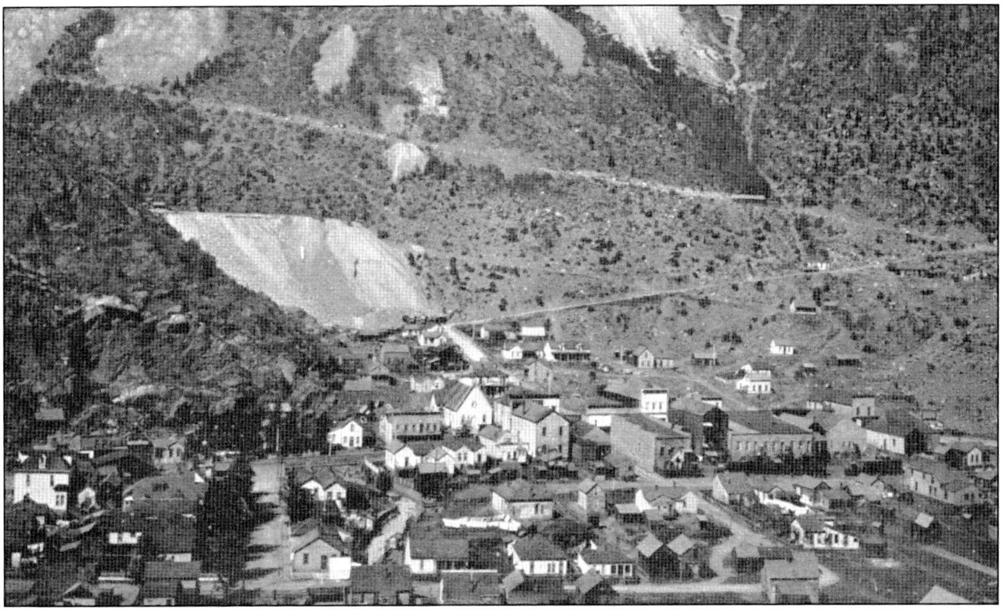

Silver Plume is shown in the late 1880s. There are many mines that were major producers in this view. The Diamond dominates the center of the photograph, and the Payrock and Dunderburg are among the other visible mines (upper right). The main business district can be seen in front of the church in the photograph's center. (USGS.)

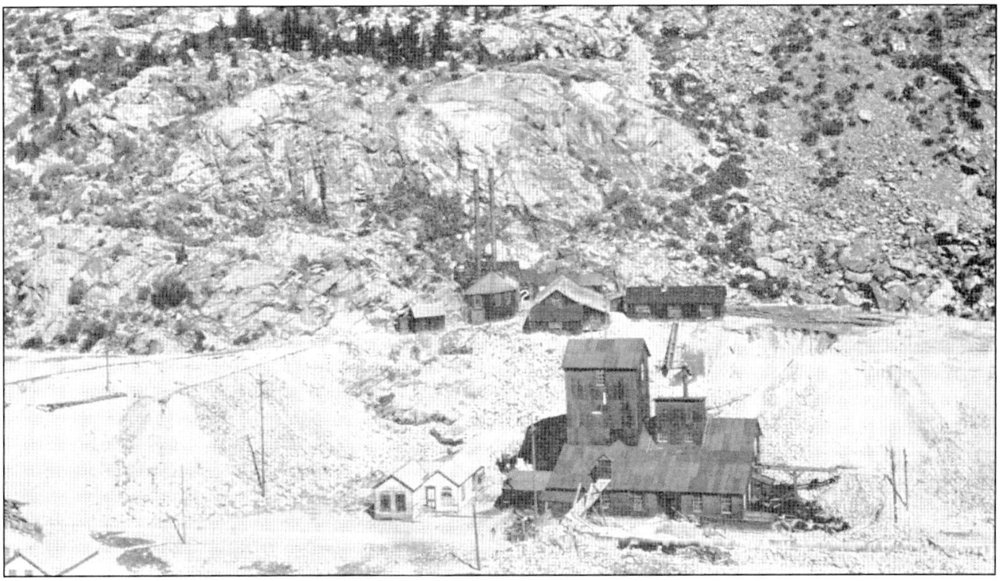

The Mendota Mill and Victoria Tunnel are pictured here at Silver Plume. Tunneling into a mountainside with the intent of intersecting other mines under the mountain was thought to be a very lucrative endeavor and explains the large number of gold mining and tunneling companies that attempted to, and did, raise capital for such an enterprise. Tunnels also served as drainage as well as a passage for pulling ore out of the mountain from other mines. Clear Creek had a fair amount of these tunnels. The Sutro Tunnel in Virginia City, Nevada, is another famous tunnel. Adolph Sutro was able to raise $3 million and later created the Sutro Baths in San Francisco, serving as mayor from 1895 to 1897. (CCHS.)

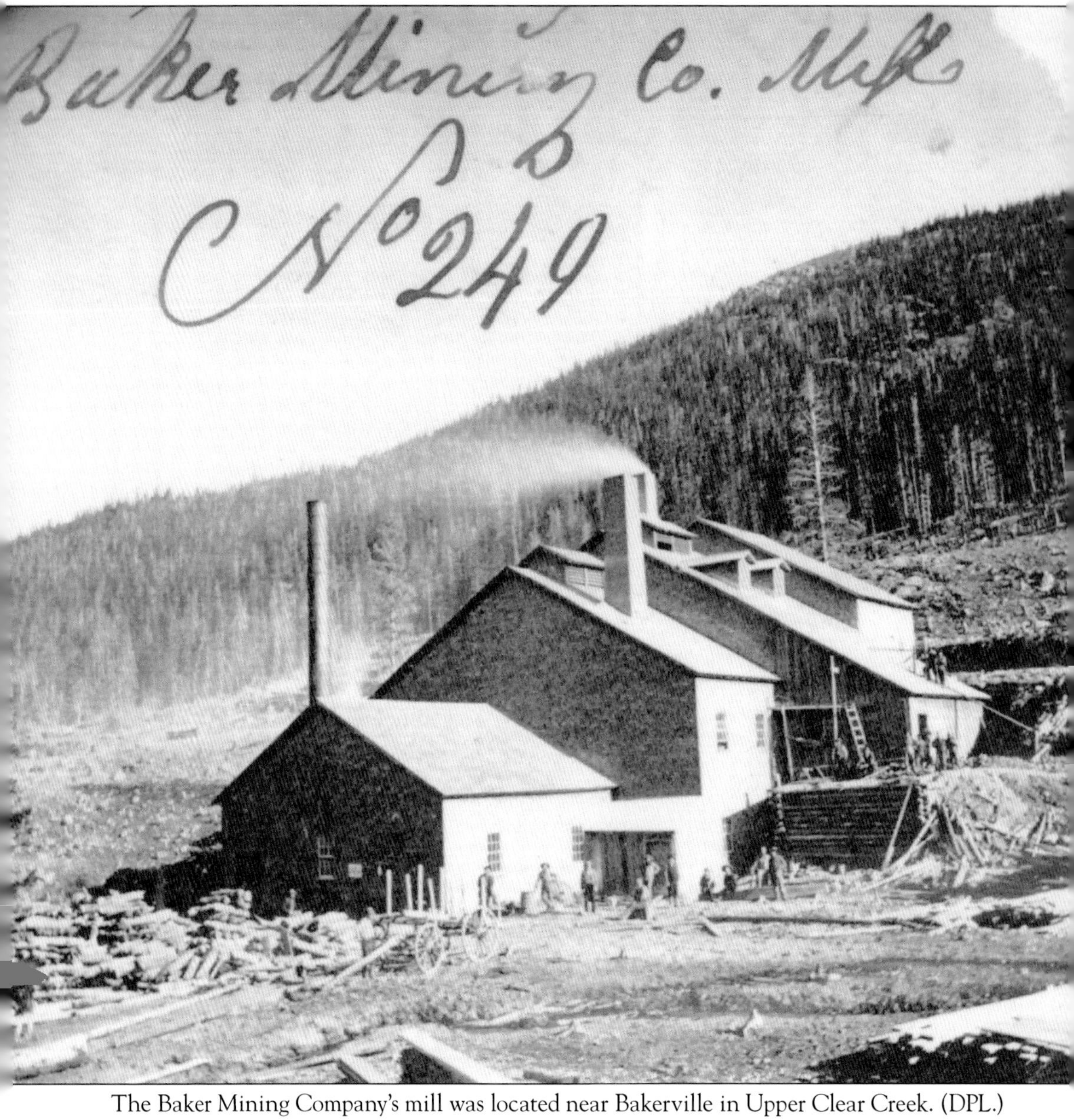

The Baker Mining Company's mill was located near Bakerville in Upper Clear Creek. (DPL.)

The Clifford Griffin monument is located above Silver Plume. Griffin worked at the Seven-thirty Mine and was distraught over a lost love. He would play the violin high above Silver Plume and, when he shot himself there in 1887, he requested to be buried at this spot. This monument is visible from the highway and town of Silver Plume to this day. The monument reads, "Clifford Griffin, Son of Alfred Griffin Esq. of Brand Hull, Shropshire, England, Born July 2, 1847, Died June 19, 1887 and in Consideration of His Own Request Buried Here." (Public Service Co.)

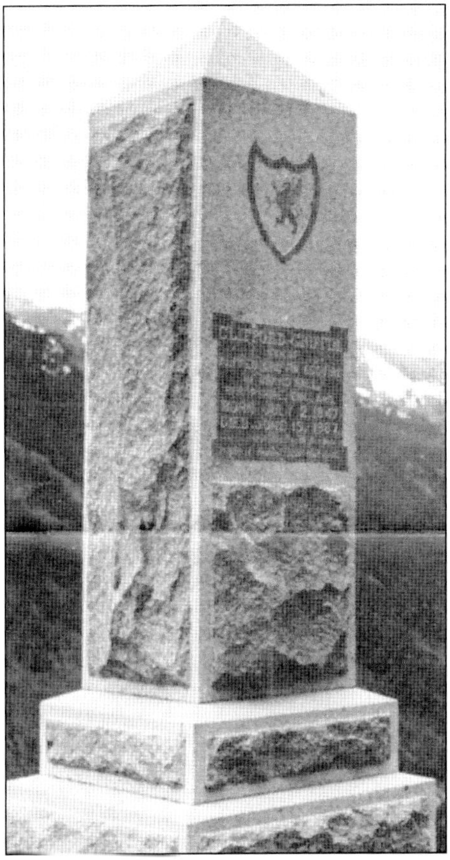

Here are the ruins of the Burleigh Mill at Silver Plume. Old mine and mill sites, such as this, were abandoned for years, and many have reached a dilapidated state. The wood rots and collapses, and time marches on. (CCHS.)

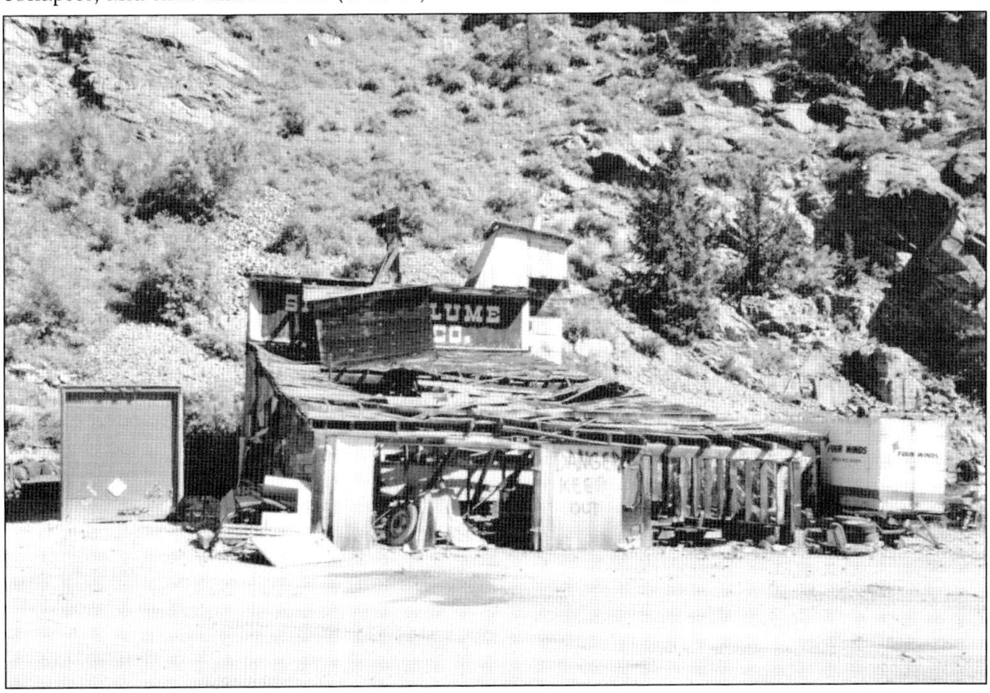

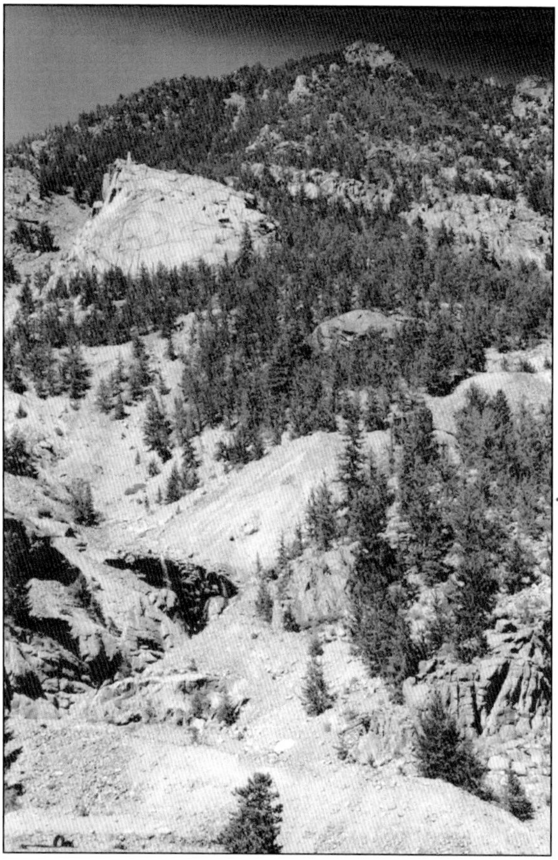

Brown Gulch west of Silver Plume shows tailings from the Seven-thirty Mine, above, and Terrible Mine, below. The Griffin Monument is visible near the top left in the photograph. (CCHS.)

Taken from the Stevens Mine, this view looks down the valley. Grays and Torreys Peaks are located to the left of the photograph, and the crest of the Continental Divide is up behind the photographer. For reference and placement, compare to the top photographs on pages 78 and 80. (CCHS.)

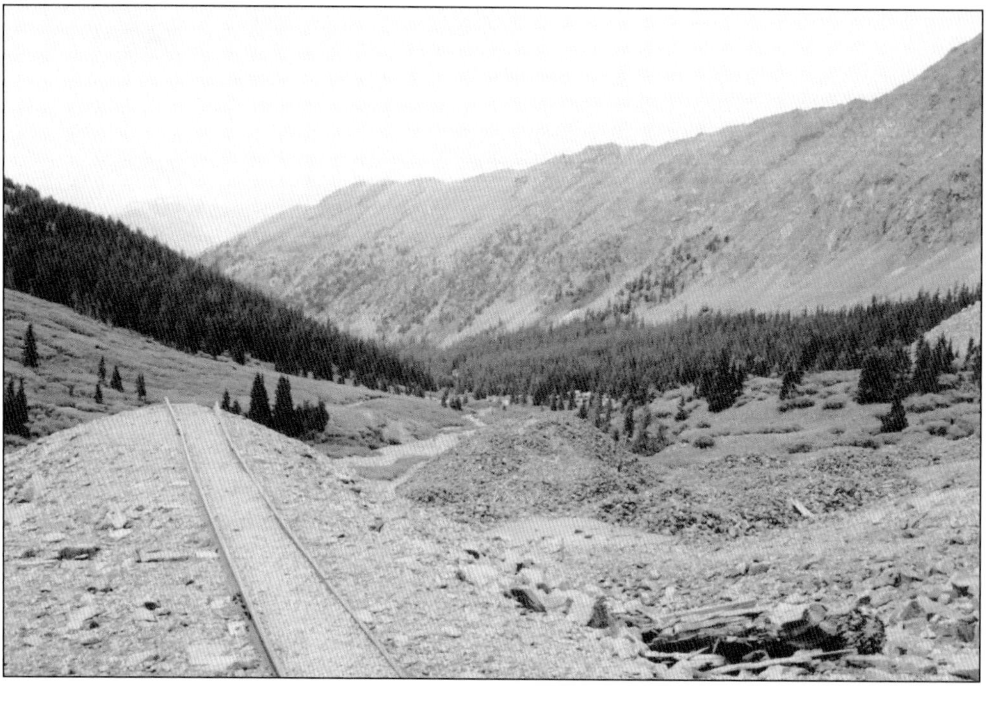

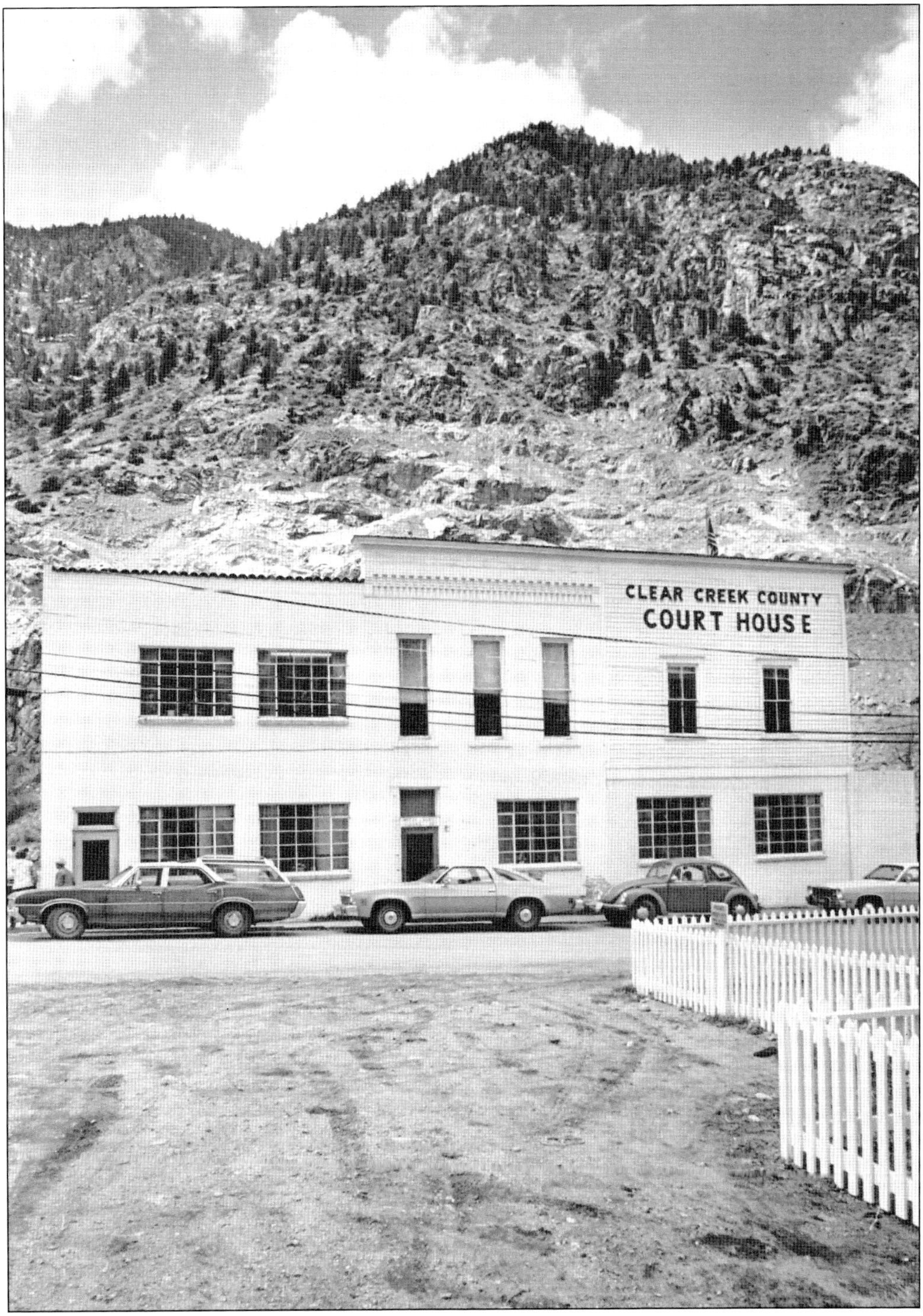

This view captures Clear Creek County Courthouse at Georgetown in 1976. Republican Mountain is in the background. The two buildings on the left were torn down and did not survive. The old courthouse was moved over a block and now serves as a community center. (*Denver Post* photograph, CCHS.)

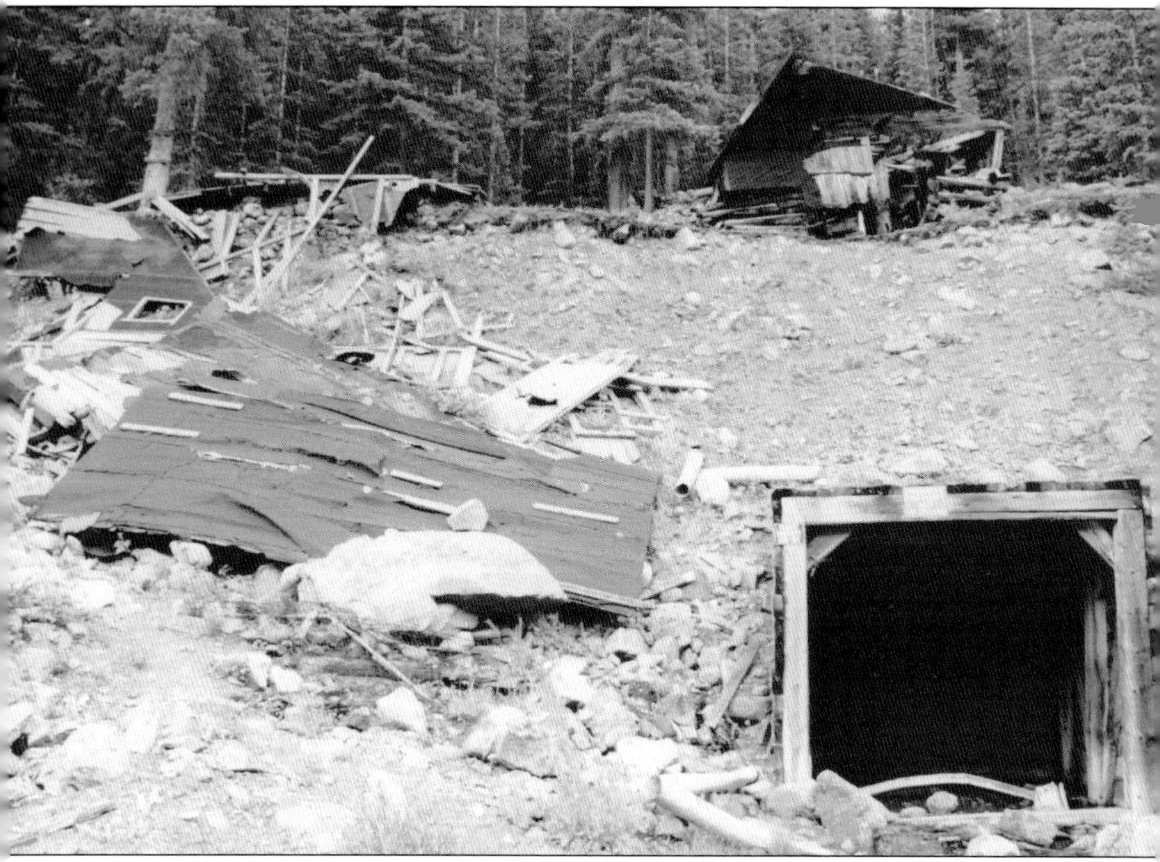

Here are the ruins of the Josephine Mine above Bakerville. In later years in the 1960s, miners would still work the old workings and truck the ore over to the smelter in Leadville. Longtime Clear Creek resident Bill Menzies was involved in the transportation of these ores. He also helped Doug Marshall work the old Santiago tailings at Waldorf in the mid-1990s. (CCHS.)

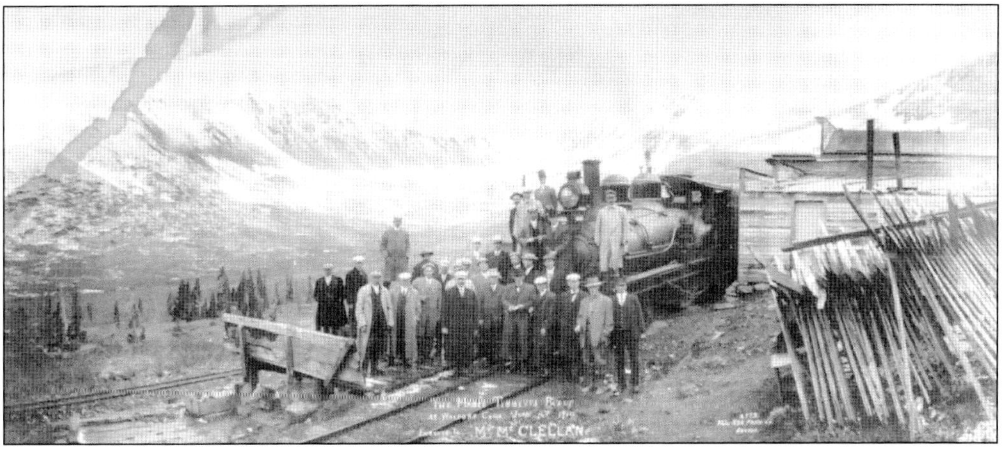

Here is the Mabee Tibbetts party at Waldorf in 1910. The Argentine Central Railroad carried these passengers all the way to the summit of Mount McClellan. The railroad was built to service the mines of Waldorf but also served as a transporter of tourists. Ed Wilcox built the line and refused to operate on Sundays. The line eventually fell into receivership due to declining ores and mismanagement. (DPL.)

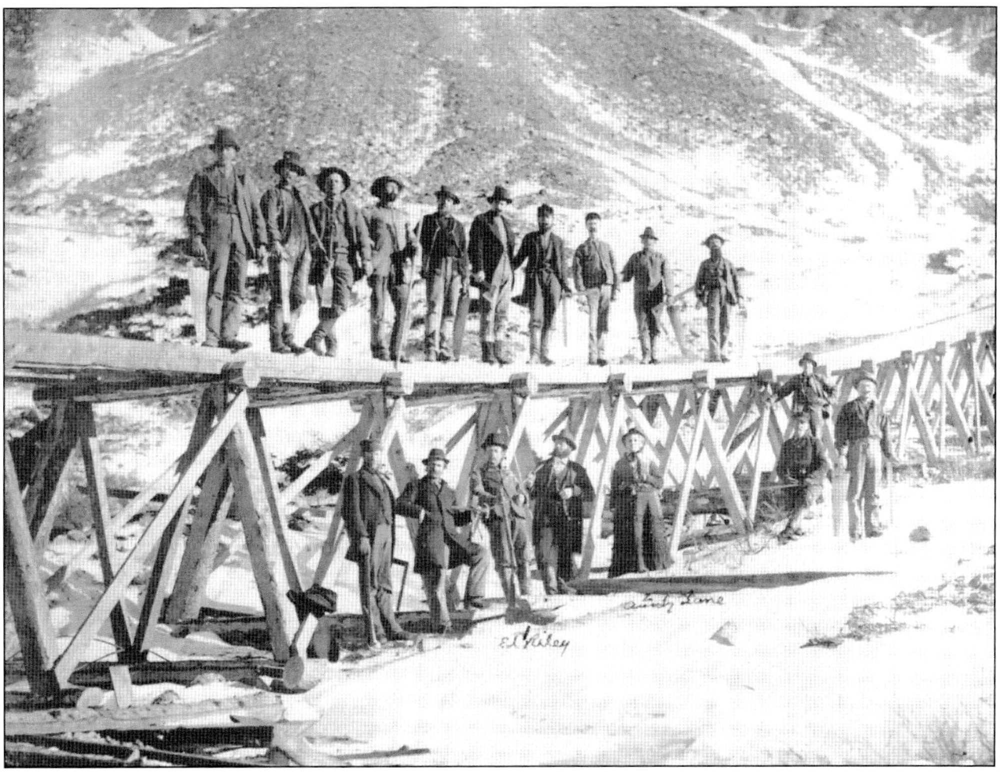

The crew of the Stevens Mine poses on the flume for a picture. The Stevens Mine was in a remote location, and its ores were transported down the valley to the Baker Co. Mill. (DPL.)

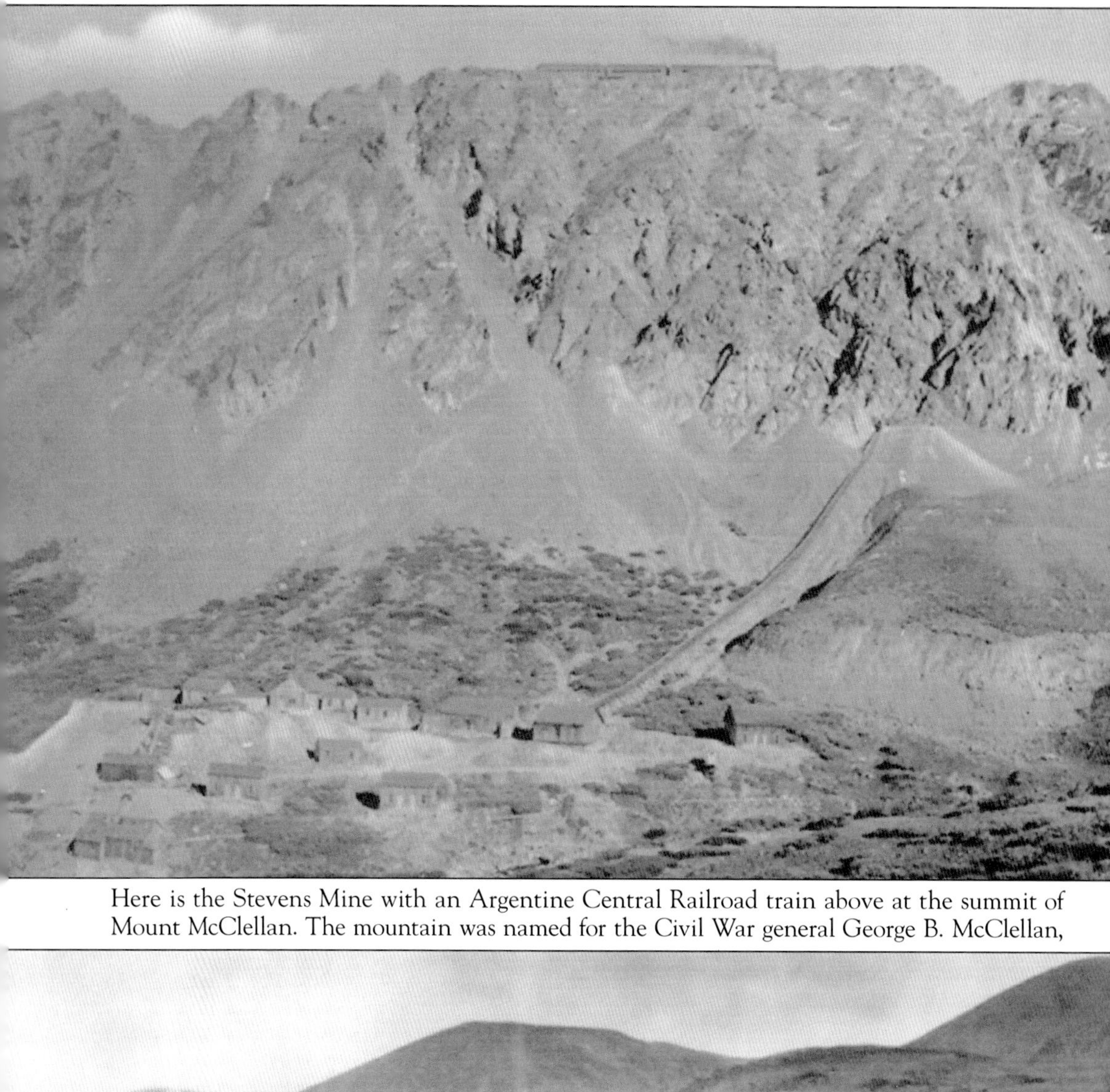

Here is the Stevens Mine with an Argentine Central Railroad train above at the summit of Mount McClellan. The mountain was named for the Civil War general George B. McClellan,

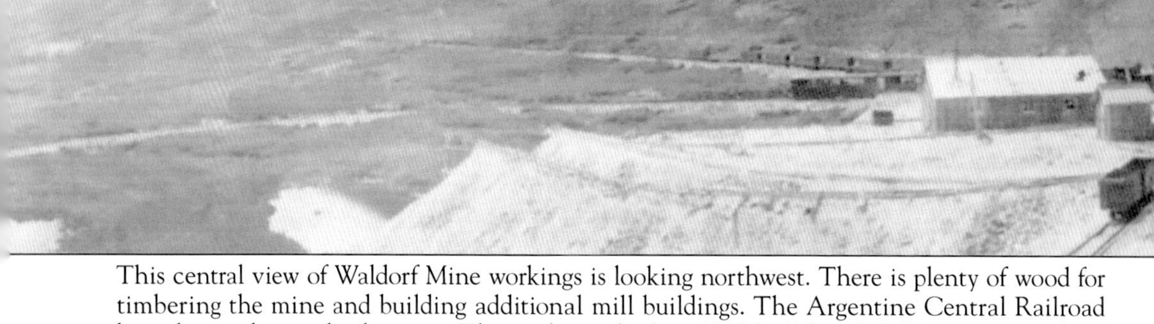

This central view of Waldorf Mine workings is looking northwest. There is plenty of wood for timbering the mine and building additional mill buildings. The Argentine Central Railroad brought supplies to this location. The grade was built in 1905 by Edward Wilcox to access these

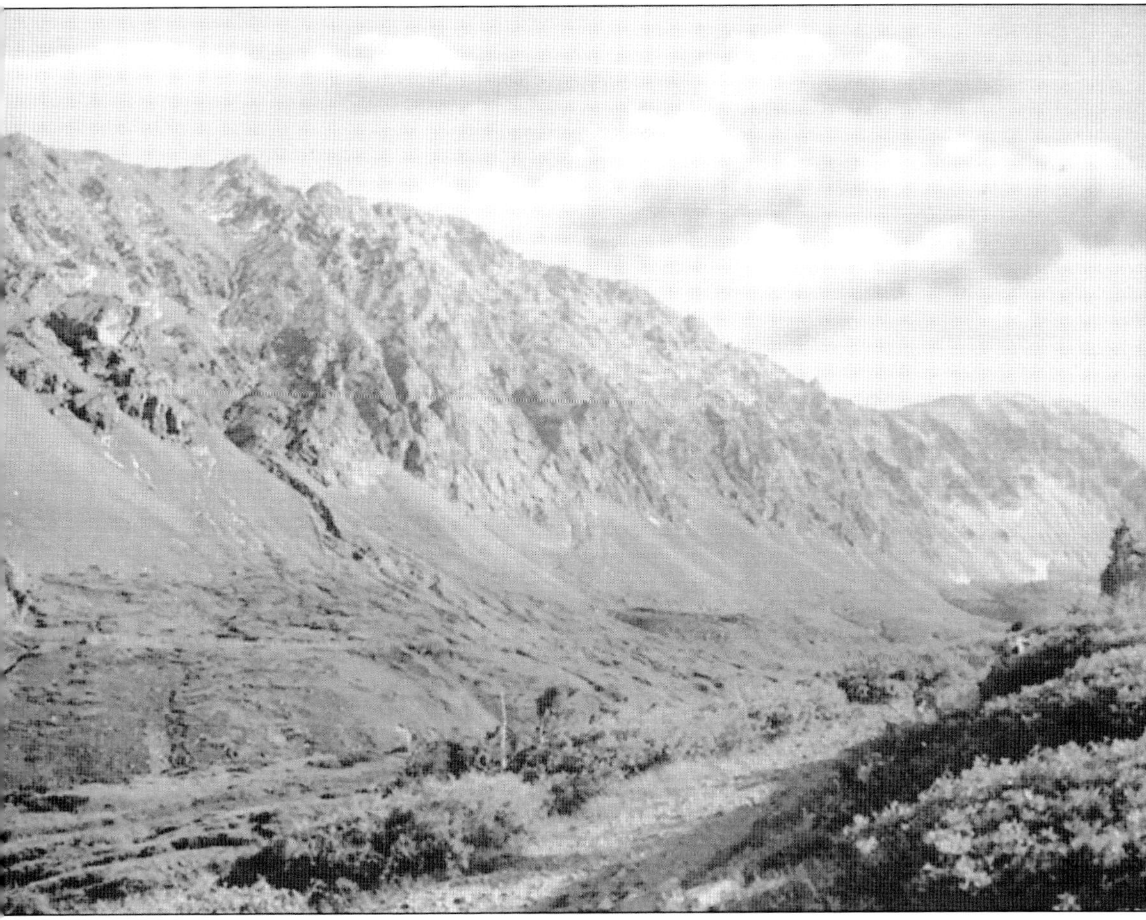

who was active in 1862 marching on Richmond. The Stevens Mine was discovered in 1862 and was a major producer for many years. (DPL.)

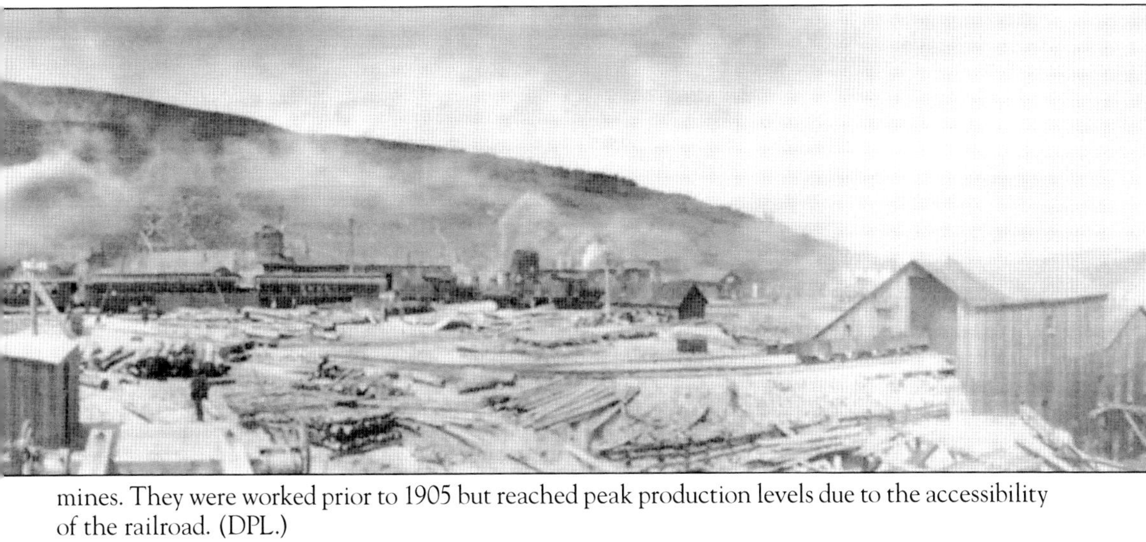

mines. They were worked prior to 1905 but reached peak production levels due to the accessibility of the railroad. (DPL.)

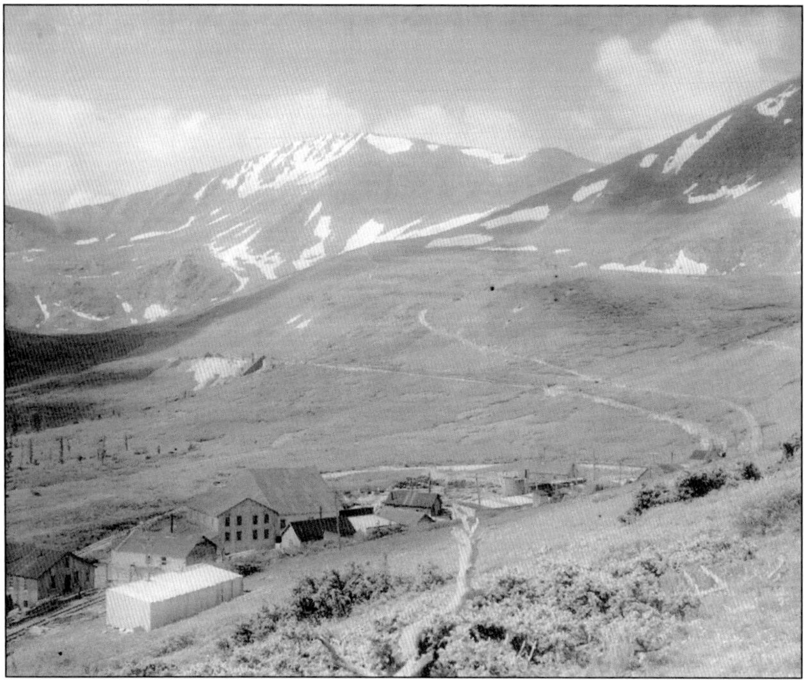

Here is a view of Waldorf and the railroad grade to the Vidler Tunnel. Argentine Pass veers off to the right and goes over the top to Montezuma and the Snake River mining district in Summit County. (DPL.)

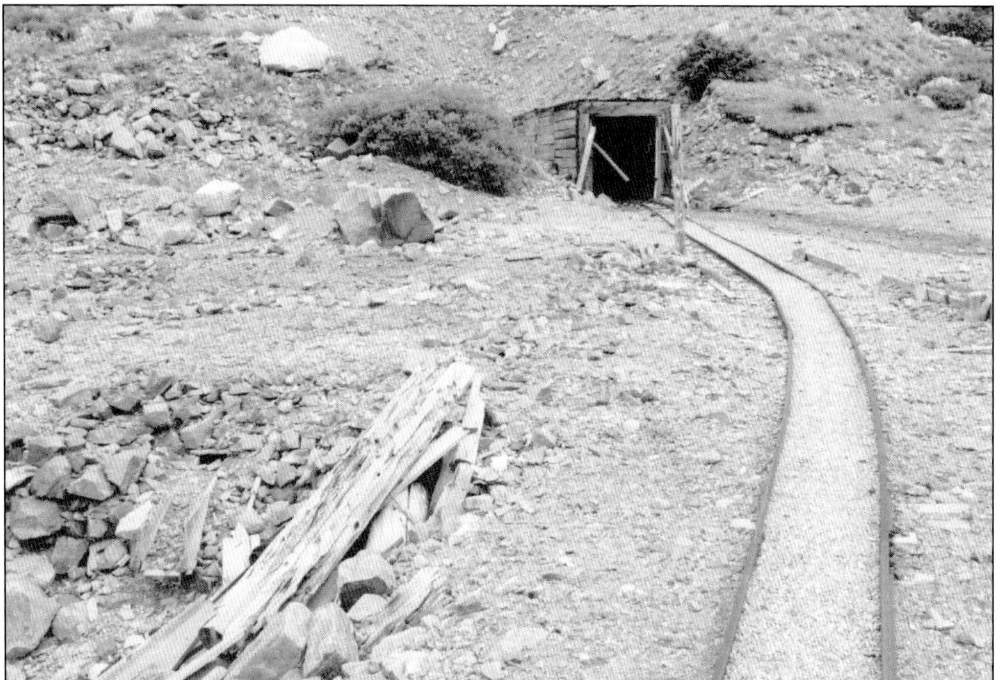

Pictured here is the long abandoned tunnel entrance at Stevens Mine near Grays and Torreys Peaks. (CCHS.)

Here is the Barton House with the annex across the street in 1873. Grace Episcopal Church, built in 1867, is seen in the lower right; a sudden mountain zephyr had torn off its roof in 1868. (USGS.)

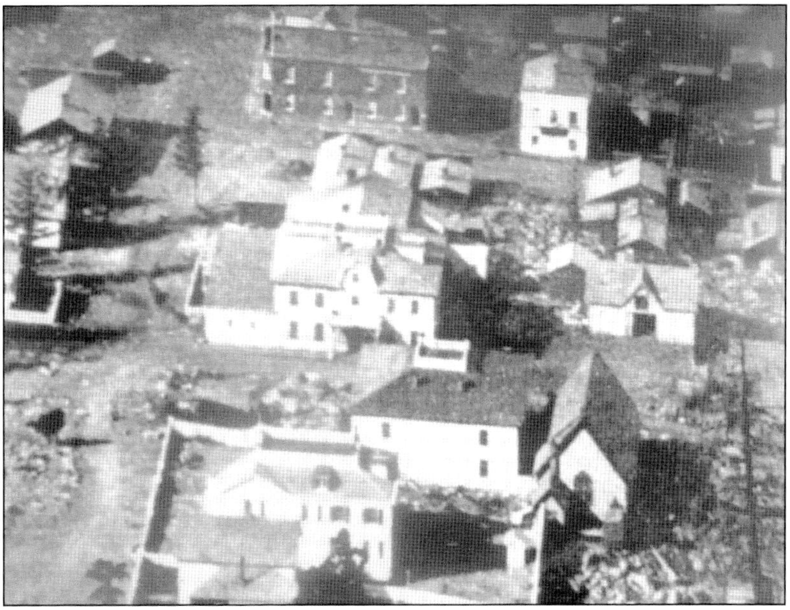

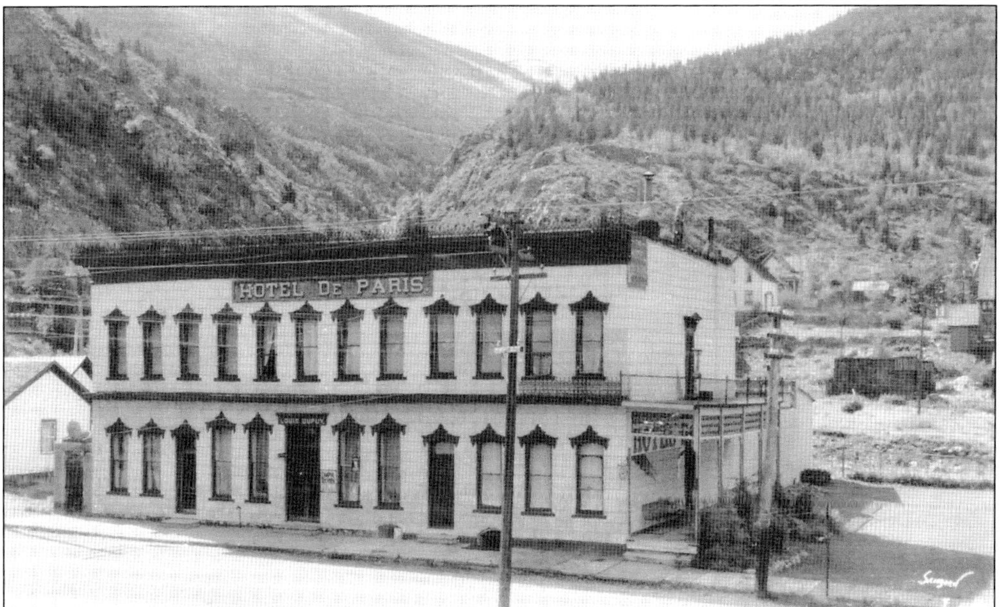

Louis Dupuy, who had suffered a mining accident in Silver Plume, built the Hotel De Paris in 1877. Following the injury, he bought the old Delmonico bakery and redesigned the building into the fanciful Hotel De Paris. It was famous worldwide for its cuisine and lodgings. Sophie Galley assisted him in his efforts, and when he died in 1900, he willed all of his possessions to her. They are both buried under a single headstone in Alvarado Cemetery. (CCHS.)

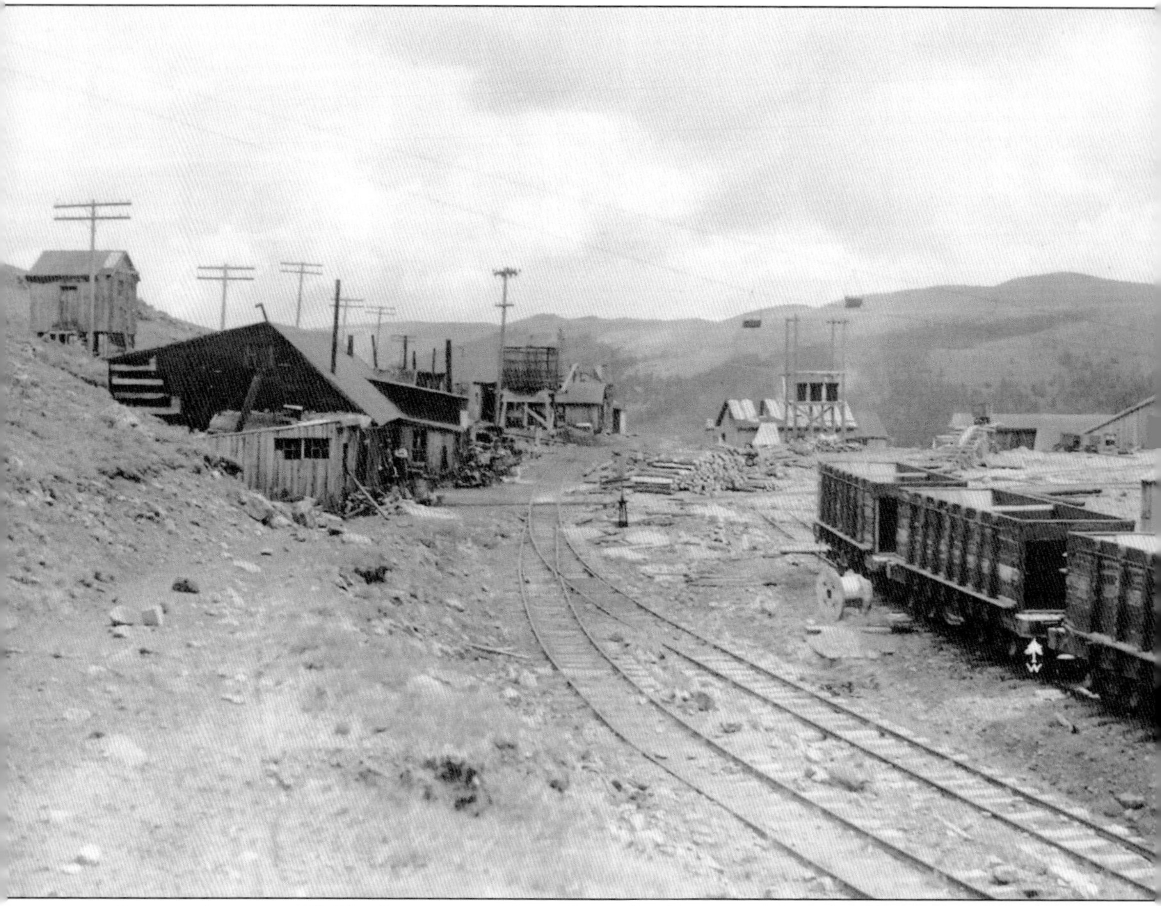

An aerial tram at Waldorf carries ore down from the Santiago Mine for delivery to the railroad. A mill was later constructed above to process the ore at that site. The tram then carried ore from the Santiago Mill. There were several mines at Waldorf including the Vidler, the Santiago, and the Waldorf. (DPL.)

Four

Empire and North Empire

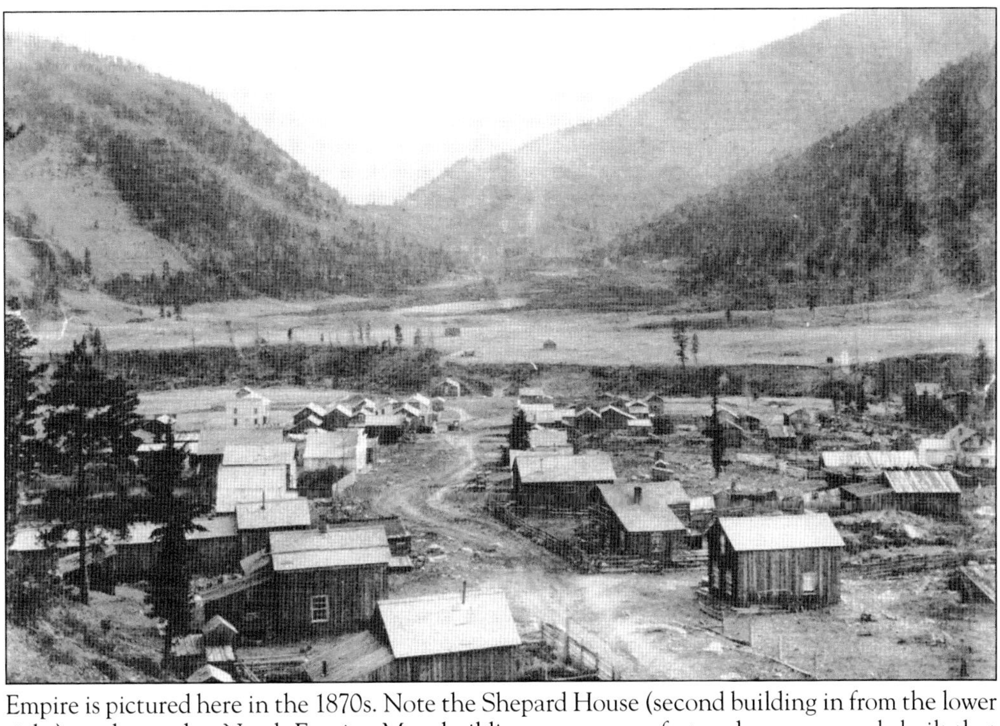

Empire is pictured here in the 1870s. Note the Shepard House (second building in from the lower right) on the road to North Empire. Most buildings went up so fast and were so poorly built that next to none remain today.

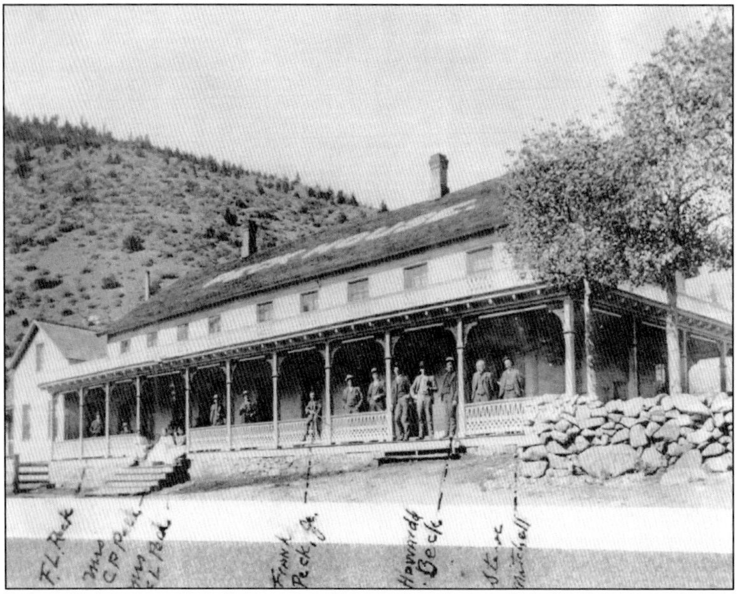

Frank L. Peck came from Chicago with his father before bringing the family out west. He and his father, James Peck, built a small home that was later expanded into a hotel to serve visiting potential investors to the Empire gold mines. James Peck died following a stagecoach accident traveling over Empire Pass in 1881 when his horse threw him. (Gary and Sally St. Clair of the Peck House.)

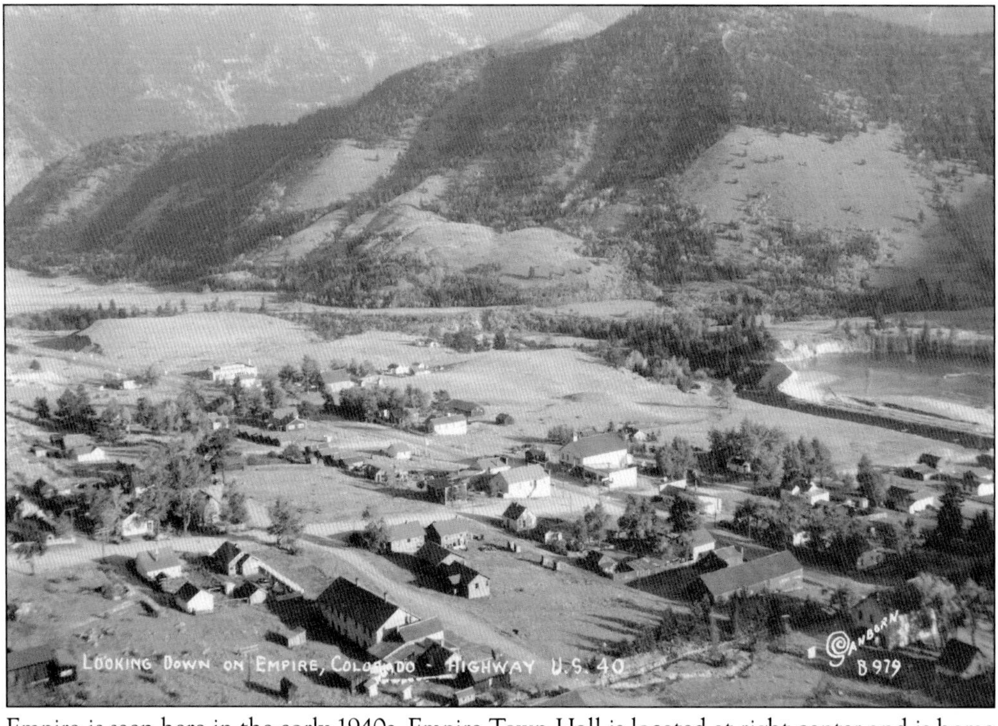

Empire is seen here in the early 1940s. Empire Town Hall is located at right center and is home to the Original Hard Rock Café. The Minnesota Mines operation in North Empire was formed when the price of gold increased to $35 an ounce in 1934. The operation utilized cyanide and created a tailings flume that transported large amounts of mine tailings to an area below Empire. The tailings pond can be seen at right center. The Peck House can be seen at lower center, and the new Empire School, built in 1938, can be seen at left center. (CCHS.)

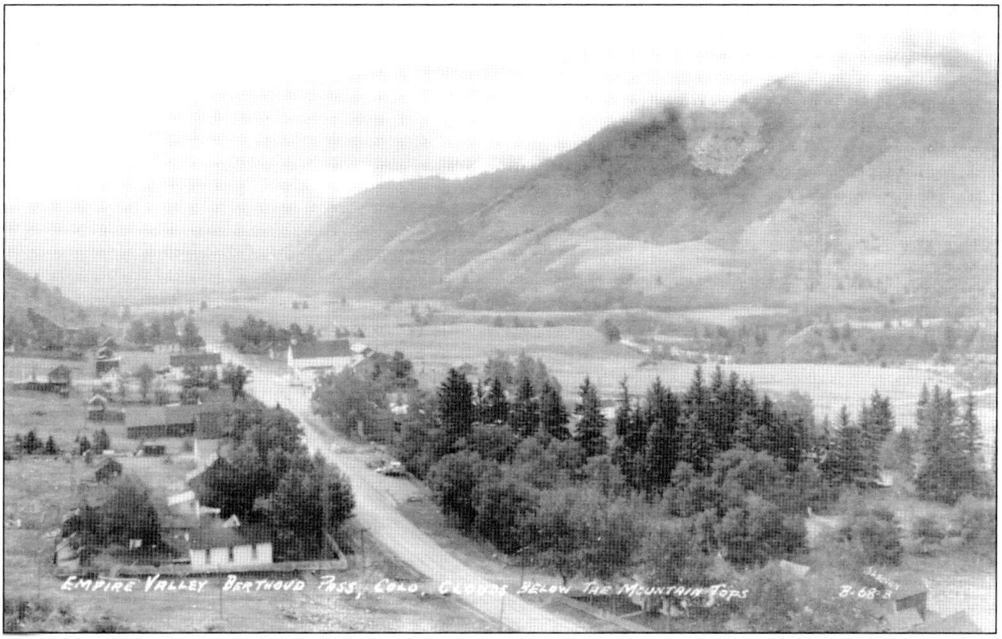

This photograph of Empire was taken in the early 1920s. It shows the landscape before the Minnesota Mines transformed the area into a tailings lake (right center). The Mint Lode ore bins can be seen on the left in the photograph, and the Badger Mill can be seen in lower right of. The Badger Mill was constructed in 1898 and failed almost immediately. (Sanborn photograph, CCHS.)

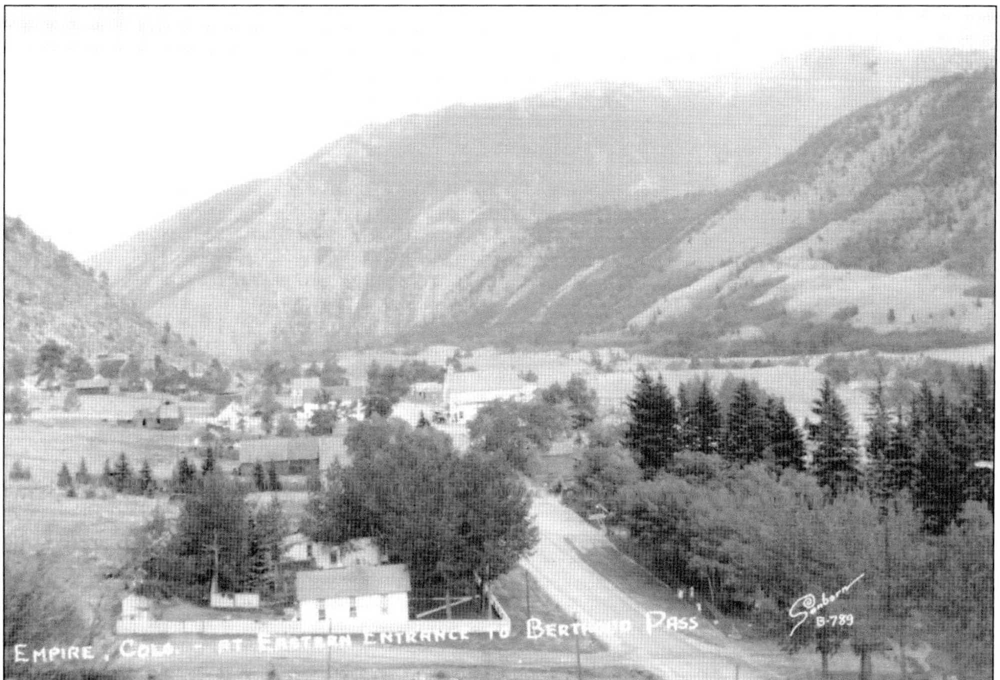

Empire is seen here in the 1930s. Note the workings of the Mint Lode (left center) and the growth of the trees along Highway 40. (Sanborn photograph, CCHS.)

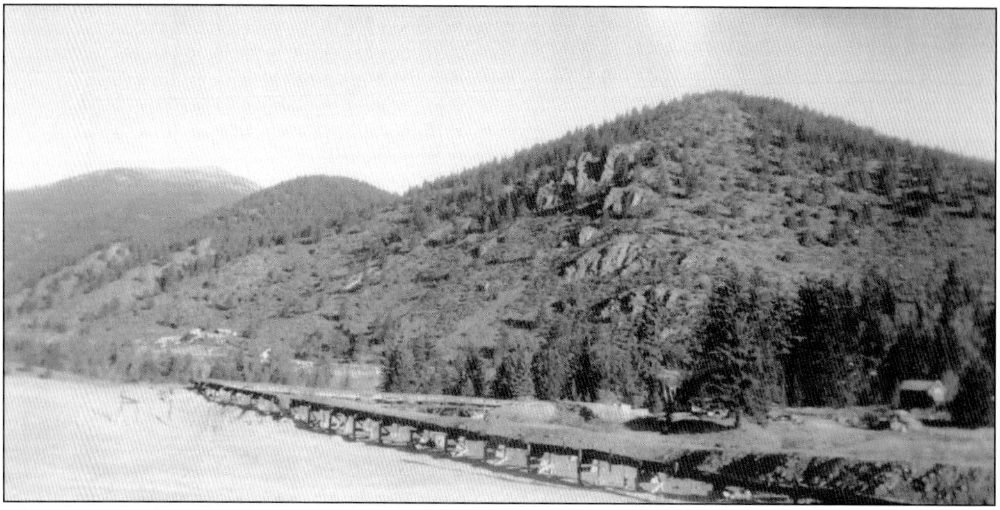

This is a view of the tailings area and flume originating from the Minnesota Mines. The author played and walked these abandoned flumes in his youth. No trace of the flume remains today. (DPL.)

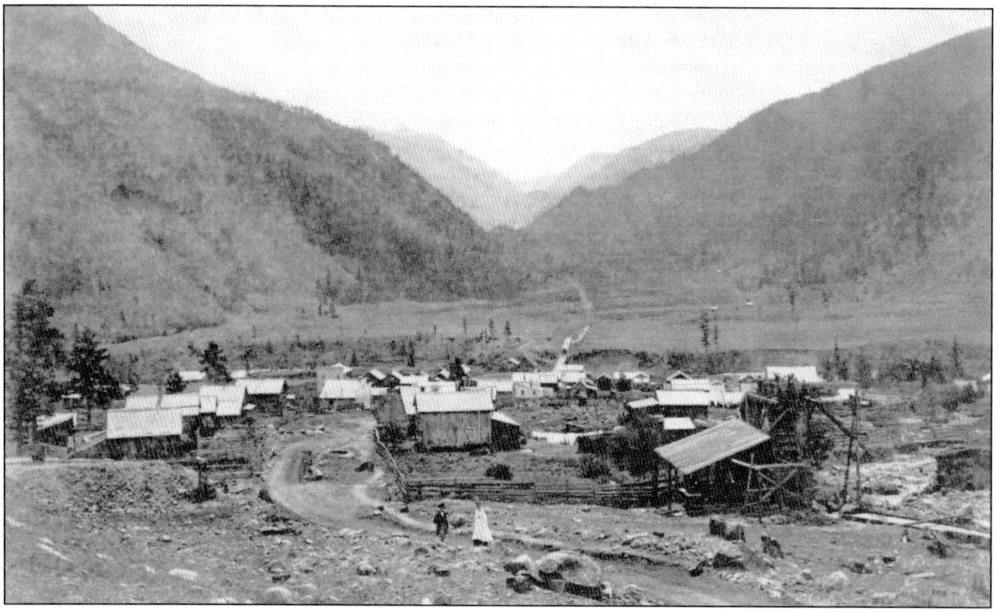

This view of Empire, looking south, shows Atlantic Mill in the lower right. Town hall can be seen at right center, and the Peck House is visible through the trees above the Atlantic Mill on the right. Covode Mountain (upper left) looms over the small mining community. (Empire Town Hall.)

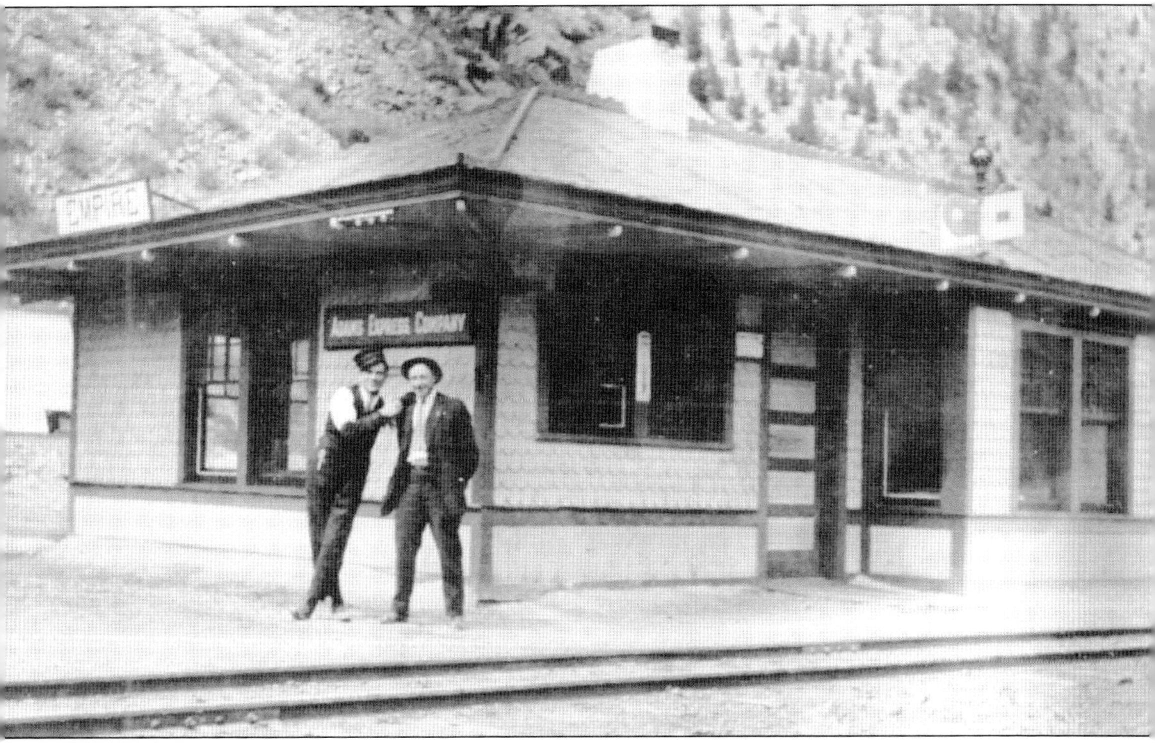

The Empire depot is pictured here in 1905. The new depot replaced the old one, which was not built close enough to the tracks. The Adams Express Company had lines all over the West, which were eventually bought by American Express. (Gary and Sally St. Clair of the Peck House.)

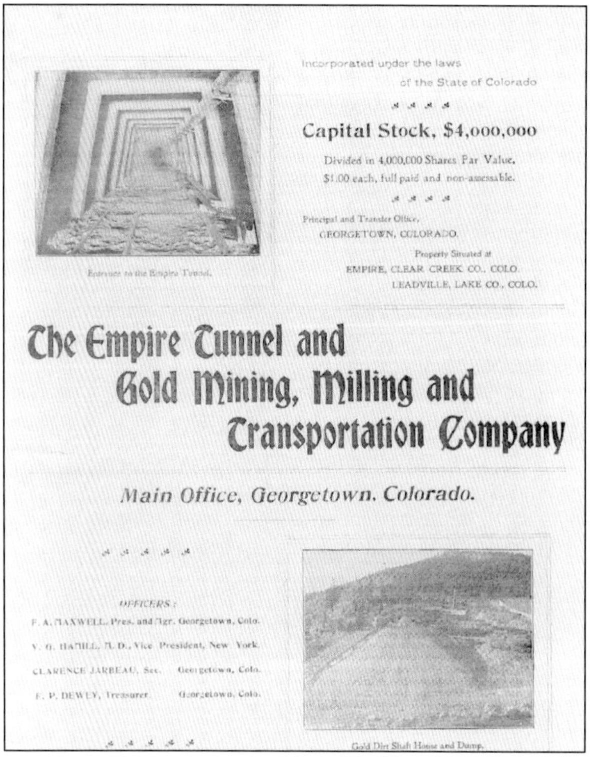

Incorporated under the laws
of the State of Colorado

Capital Stock, $4,000,000

Divided in 4,000,000 Shares Par Value,
$1.00 each, full paid and non-assessable.

Principal and Transfer Office,
GEORGETOWN, COLORADO.

Property Situated at
EMPIRE, CLEAR CREEK CO., COLO.
LEADVILLE, LAKE CO., COLO.

Entrance to the Empire Tunnel.

The Empire Tunnel and Gold Mining, Milling and Transportation Company

Main Office, Georgetown, Colorado.

OFFICERS:

F. A. MAXWELL, Pres. and Mgr. Georgetown, Colo.
V. G. HAMILL, M. D., Vice President, New York.
CLARENCE JARBEAU, Sec. Georgetown, Colo.
F. P. DEWEY, Treasurer. Georgetown, Colo.

Gold Dirt Shaft House and Dump.

The Empire Tunnel Gold Mining and Milling Company drove the Empire Tunnel at Empire Junction to the mines at North Empire. (Roger Dickinson, Empire Town Hall.)

This image shows a line of tunnel for Empire Tunnel Company. Frank Maxwell was an officer in this company and lived in the fancy Victorian house near the Barton House, now known as the Maxwell House, in Georgetown. It is the highlight of any visit to Georgetown. (Roger Dickinson, Empire Town Hall.)

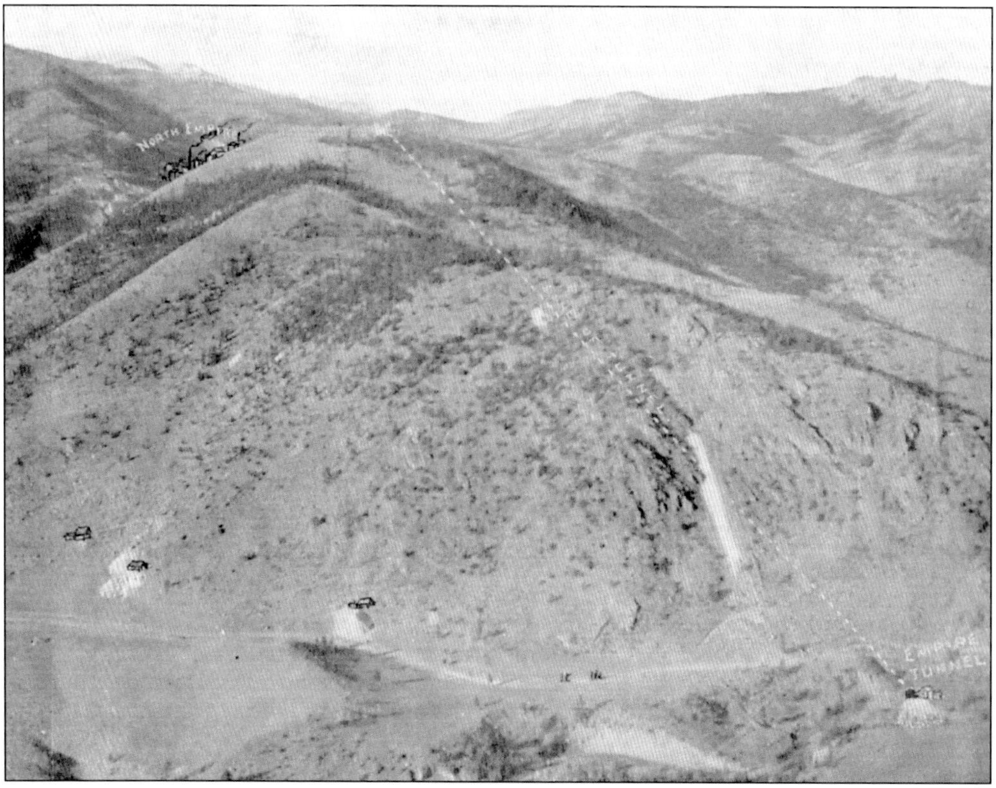

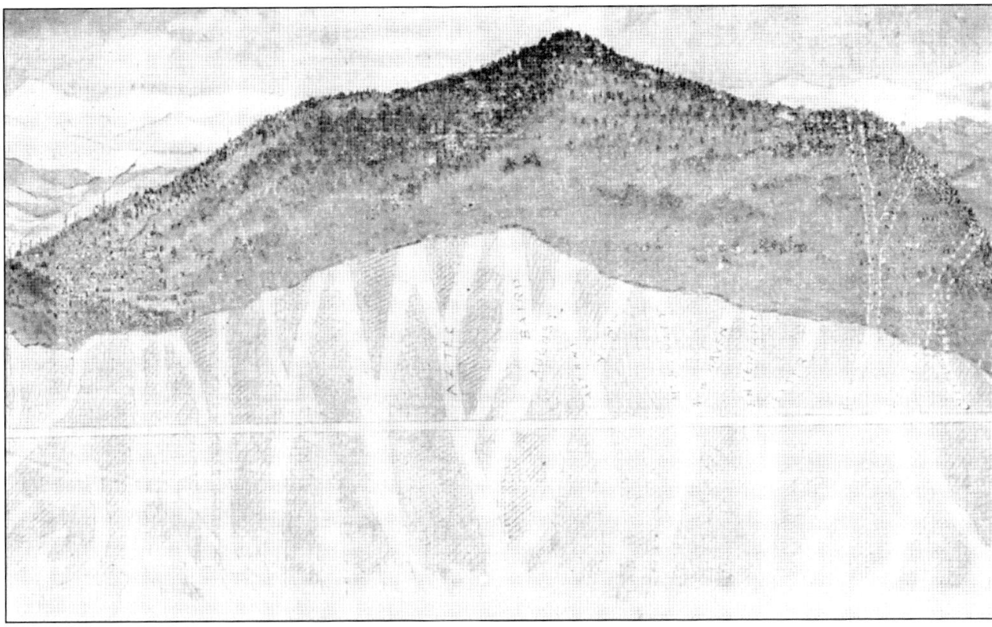

Here is a depiction of the view underneath Covode Mountain for the Empire Tunnel Gold Mining and Milling Company. The tunnel company ran into financial difficulties but, in its prime, proved to be quite lucrative for the owners and investors. (Roger Dickinson, Empire Town Hall.)

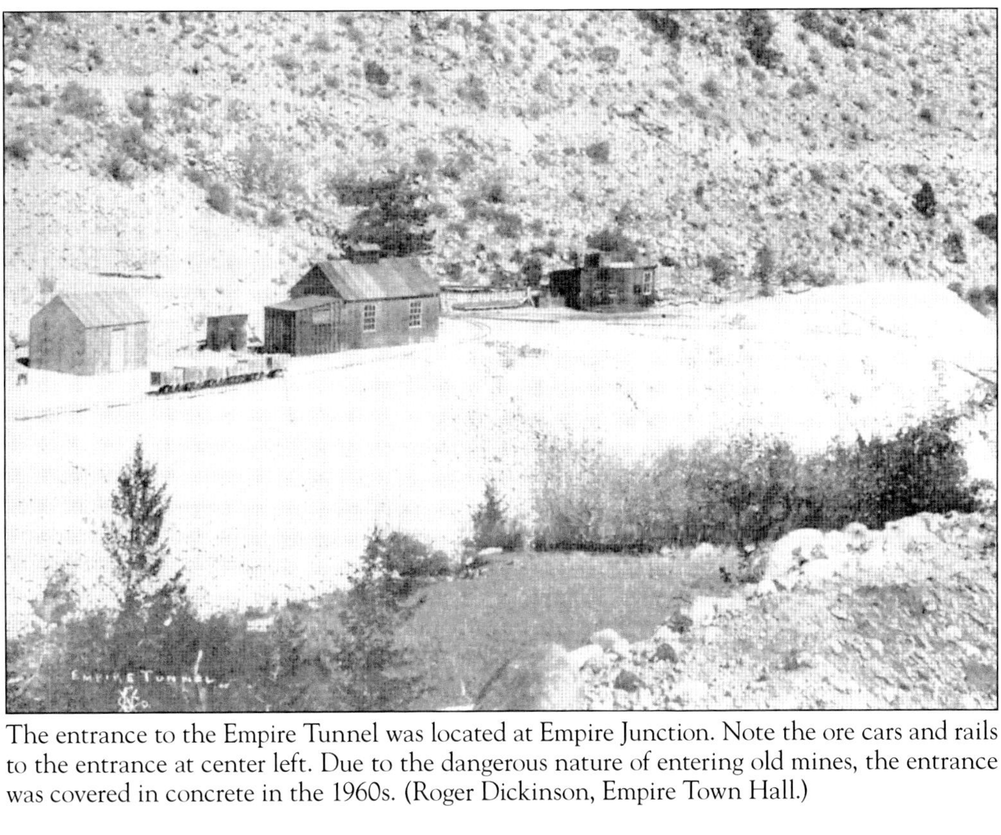

The entrance to the Empire Tunnel was located at Empire Junction. Note the ore cars and rails to the entrance at center left. Due to the dangerous nature of entering old mines, the entrance was covered in concrete in the 1960s. (Roger Dickinson, Empire Town Hall.)

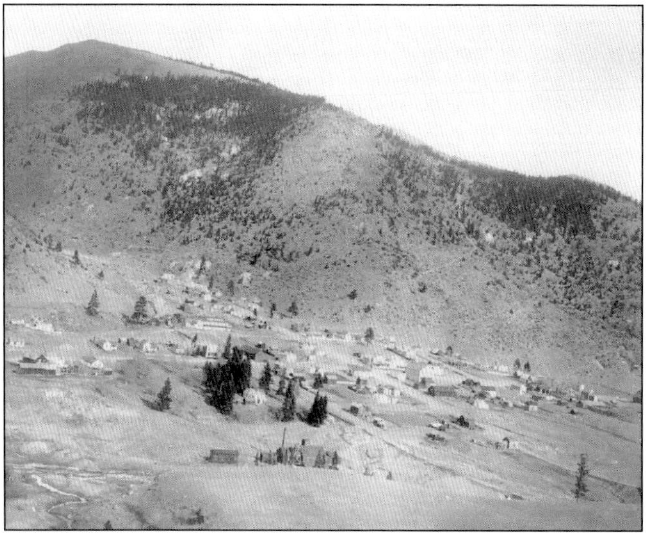

Lachlan McLean took this photograph on Lincoln Mountain in Empire around 1915. He had to carry his pioneer photograph equipment up the mountain for a good distance to get the image. The Atlantic Mill (lower left) can be seen, the Peck House is pictured at left center, and the town hall as well as the 1897 schoolhouse are at right center in the lower portion of the photograph. The Redmen hall, old Maire barn, and Atlantic Mill are also visible in this view. (UCB.)

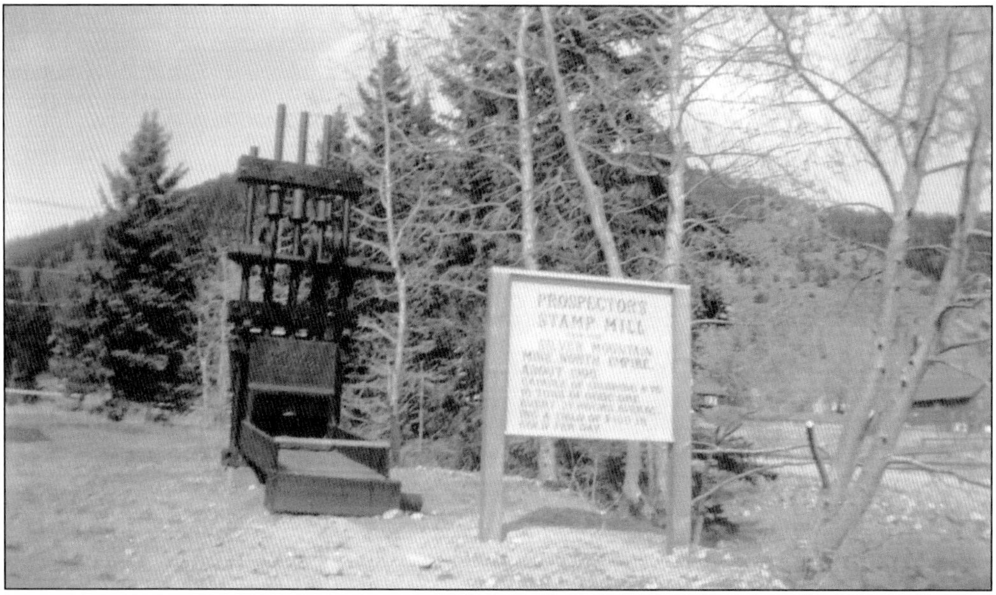

The Empire museum and library building was owned and operated by Louise C. Harrison and Margaret Collbran (granddaughters of the Coors family). The Redmen hall museum had dioramas depicting mining scenes, which were built by Hank Gentsch. Hank Gentsch also had a museum in Central City where many of his dioramas were on display for years until gambling was legalized in the early 1990s. Annie Oakley's store in Central City is where the museum was in the 1970s. The sign reads, "Prospectors Stamp Mill from the Silver Mountain Mine about 1900. Capable of crushing ore every 24 hours averaging a yield of $100 in gold per day." (DPL.)

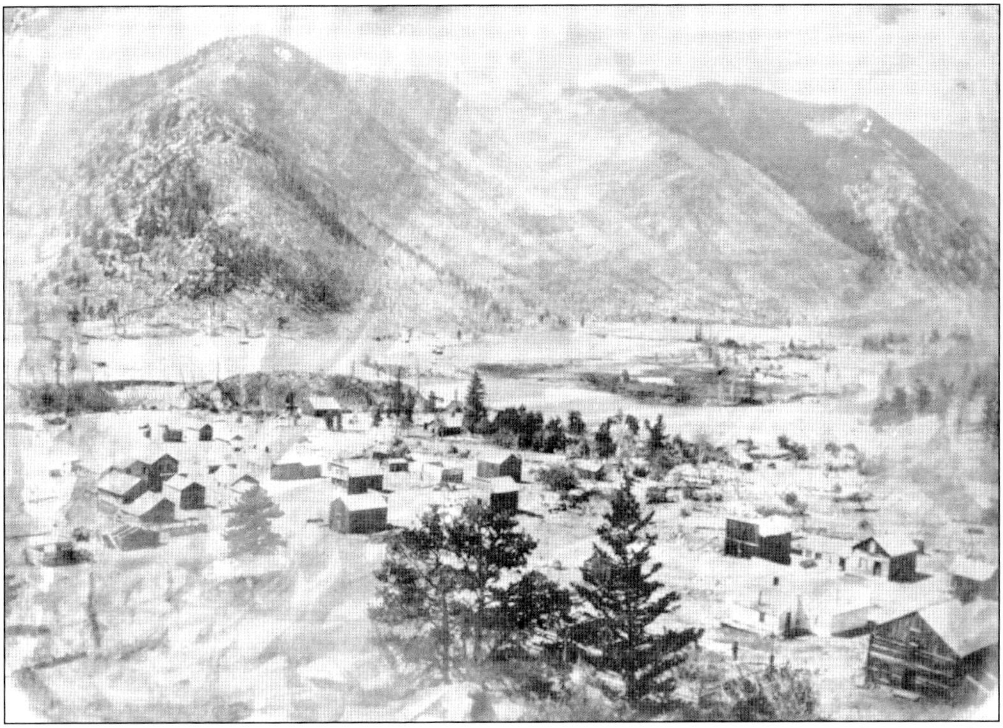

Empire is pictured here in 1863. The D.W. King house can be seen in the lower right. Few other structures remain today from this photograph. The saloon building (lower left) now houses Jenny's restaurant.

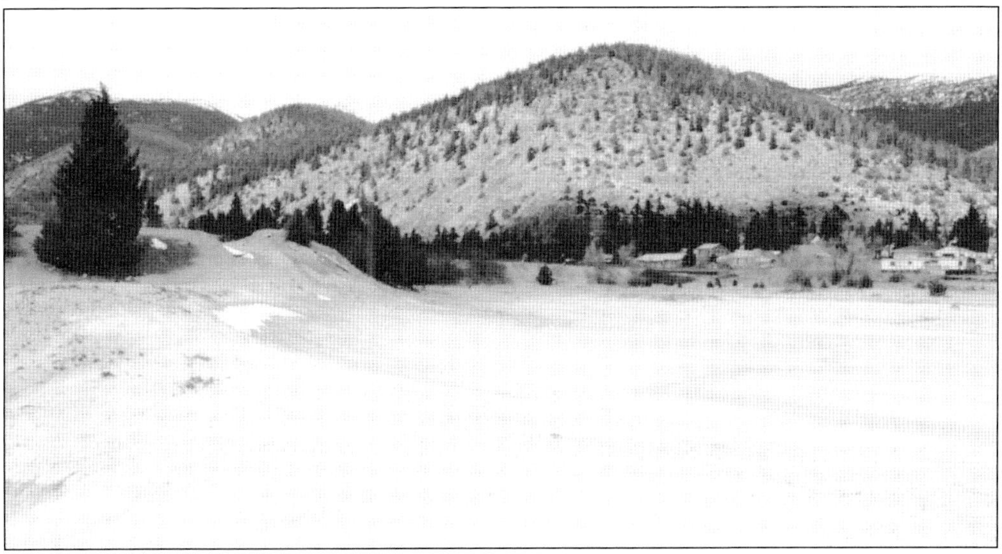

These tailings (polluted mine waste) were flumed down from the North Empire location of the Minnesota Mines. In the 1930s, when the price of gold increased to $35 due to Roosevelt's New Deal, many old mines opened up and were again gold producers. The Minnesota Mines was a result of a sizable investment from the 3M Corporation in Minneapolis. The operation became a major gold producer, leading Clear Creek County in gold production for many years in the late 1930s and early 1940s. The company built a 250-ton-per-day mill at the North Empire location. (CCHS.)

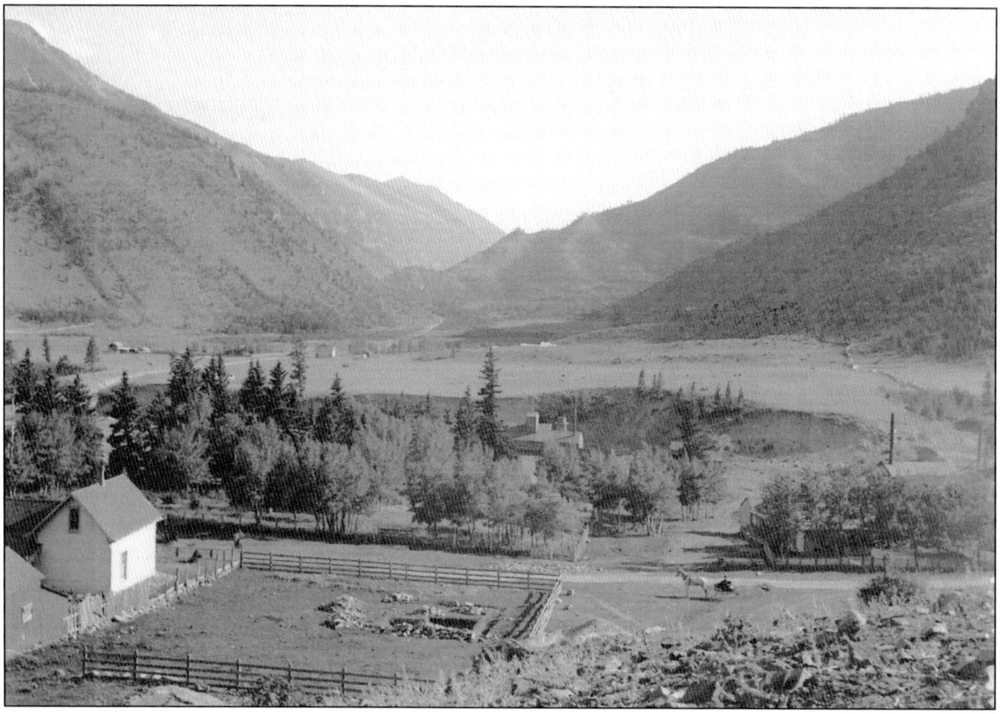

This c. 1910 view is of Empire looking south toward Georgetown. The Atlantic Mill can be seen lower center, and the Badger Mill, built in 1898, is at the lower right. The mill had been long abandoned by the time this photograph was taken.

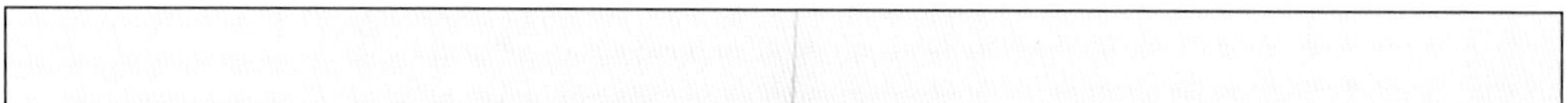

Empire and Route 40 are pictured here in 1926. The Model T is stopped across from the town hall. The saloon building, where gasoline and tourist wares were sold, is seen at right. In the distance, the Silver Creek trail grade is visible. This was built as a railroad grade back along the valley to the silver producing camp at Silver Creek, located above Lawson. (Windfall Photography, Gary and Sally St. Clair of the Peck House.)

Marshall Park is seen here in 1900. This area is located in the Montana mining district. The railroad stopped here on its way to Georgetown. It is no longer referred to as Marshall Park but Empire Junction. The Marshall and Russell Tunnel was driven here in 1900 and was extended to a distance of 2,500 feet.

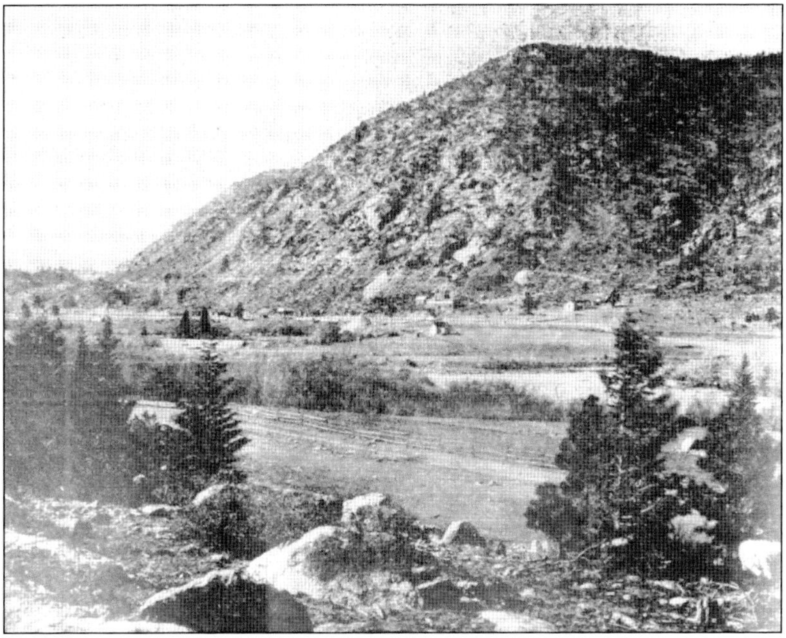

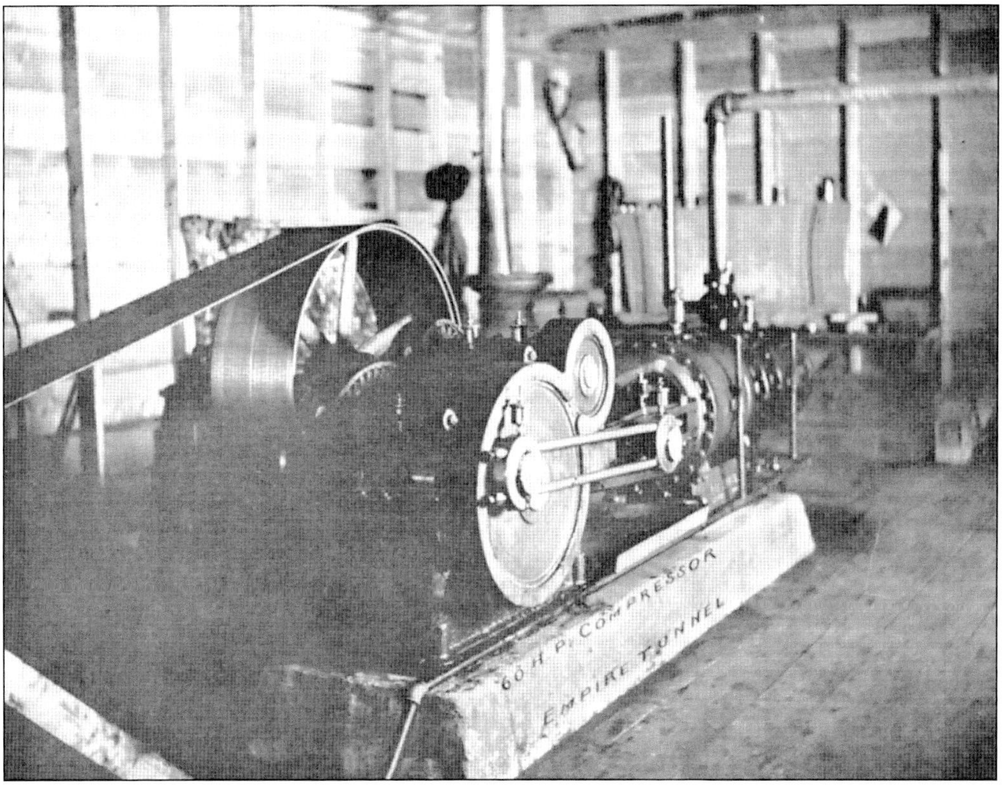

Here, a 60-horsepower air compressor is used at the Empire Tunnel. Compressed air was used for pneumatic air drills, power, and to provide fresh air for the miners deep in the tunnel. (Roger Dickinson, Empire Town Hall.)

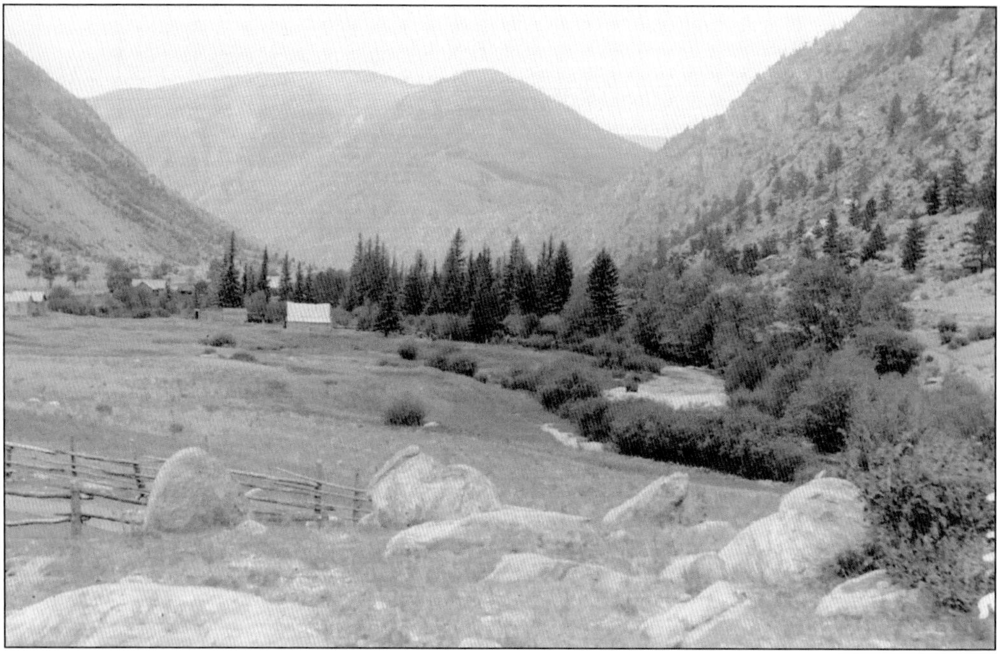

The saddleback is pictured here between Empire (right) and Georgetown (left). Empire Pass can be seen in a diagonal line from Alvarado Ranch. The railroad runs on the right, and the stage road to Georgetown runs on the left. The Orsi Ranch is located at left center. (DPL.)

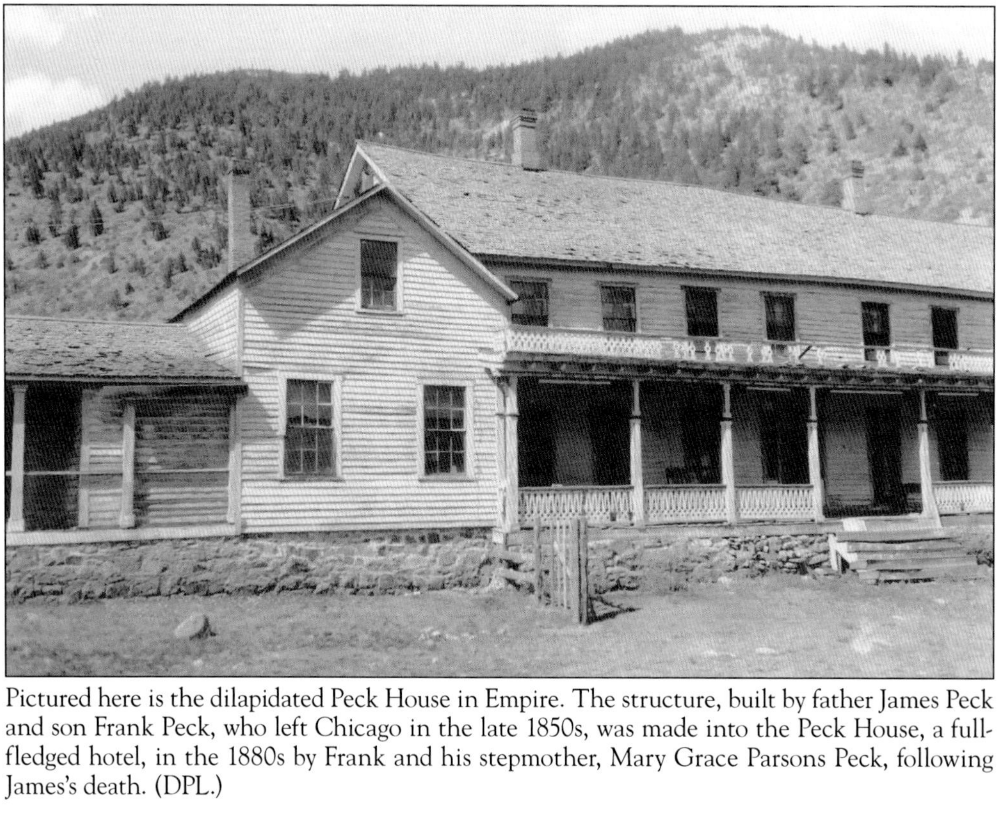

Pictured here is the dilapidated Peck House in Empire. The structure, built by father James Peck and son Frank Peck, who left Chicago in the late 1850s, was made into the Peck House, a full-fledged hotel, in the 1880s by Frank and his stepmother, Mary Grace Parsons Peck, following James's death. (DPL.)

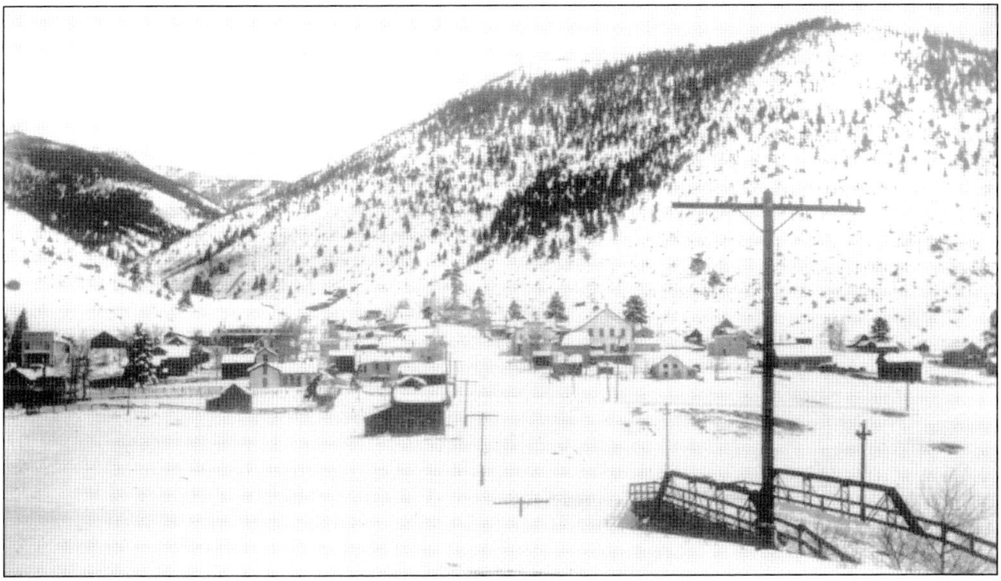

In this c. 1910 image, note the Empire Town Hall (right center), the Kistler Mill (left center), and the Redmen hall with two upstairs windows and a false front (left center). This building no longer has a false front. In 1922, the bureau of public roads needed an office to oversee the construction of the new road over Berthoud Pass. The roof had been completely destroyed by the big snow of 1913, and the building sat vacant for nine years until a new one was constructed without a false front. In 1938, the refurbished Kistler Mill ran for a short time and then burned down.

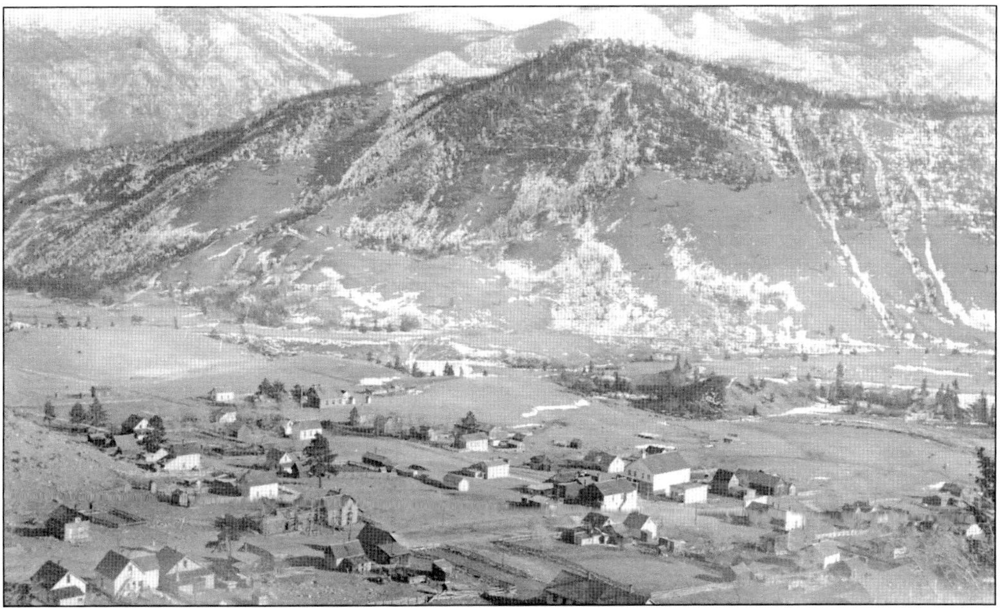

This view of Empire in 1911 looks southeast and shows the old 1897 schoolhouse (left center), town hall (lower center right), Peck House (lower bottom), D.W. King House (lower left), the piece of land that was reserved for Minnesota Mines tailings (lower right), and the protruding bluff with its cemetery (right center). In Central City, a large mining operation, the Chain 'O Mines Corporation, consolidated the mining claims in the patch and glory hole. Compare this image to the photograph on page 84.

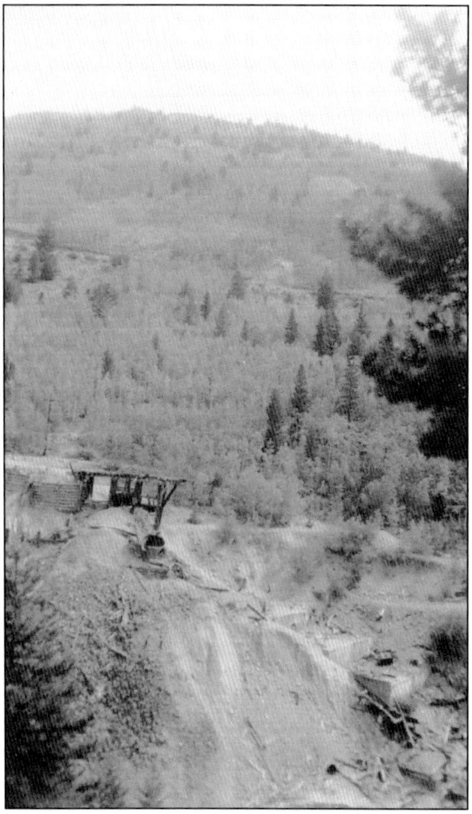

This is a view of the old trestle and Empress Mill site in North Empire. This mine was a major gold producer in North Empire and had the main trunk Aorta Tunnel.

The gold fissure was discovered early on and was one of North Empire's main producers. North Empire was also referred to as the Upper Union mining district and often confused with another Empire district formed near Idaho Springs in 1860.

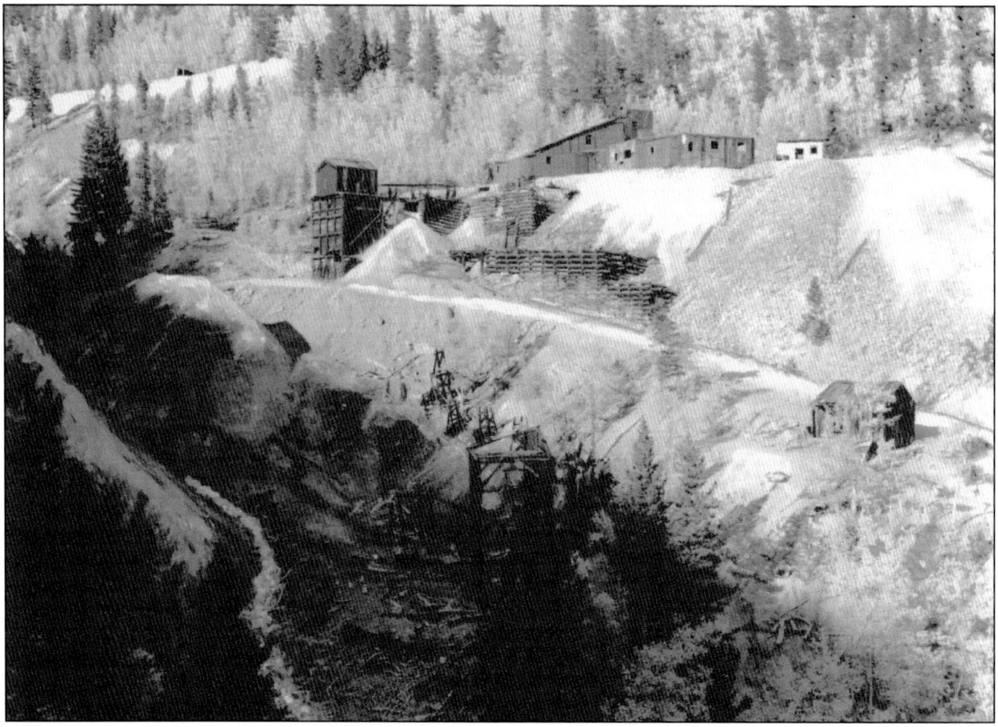

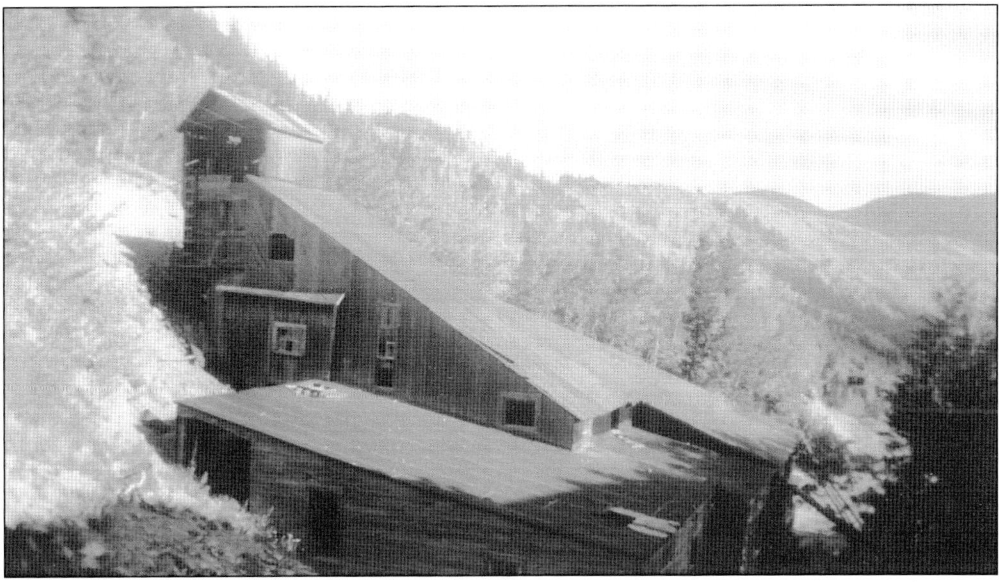

Seen here is old Conqueror Mill in North Empire. The Conqueror Mine was discovered in 1877 and was active for many years. A boardinghouse was built for the miners directly across the gulch in 1910. Hippies were using this building in the 1960s, and it is rumored the local sheriff's department burned the building down to move the hippies along.

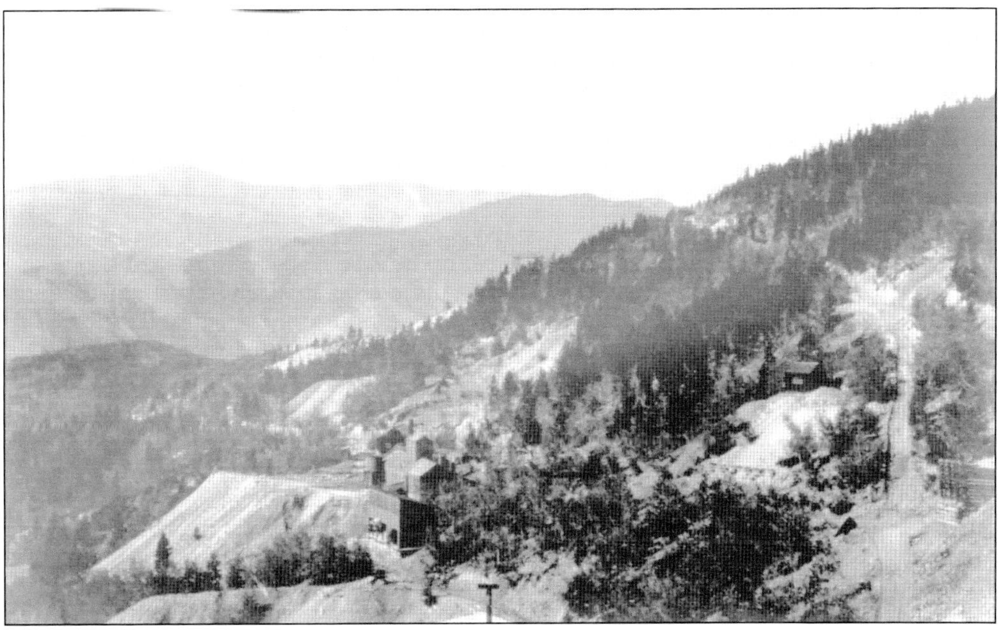

Silver Mountain Mine and Mill in North Empire is pictured here in 1934. The Tenth Legion Mine is located to the right. All of these Upper Union mines benefited and experienced new life due to the price increase of gold in 1934.

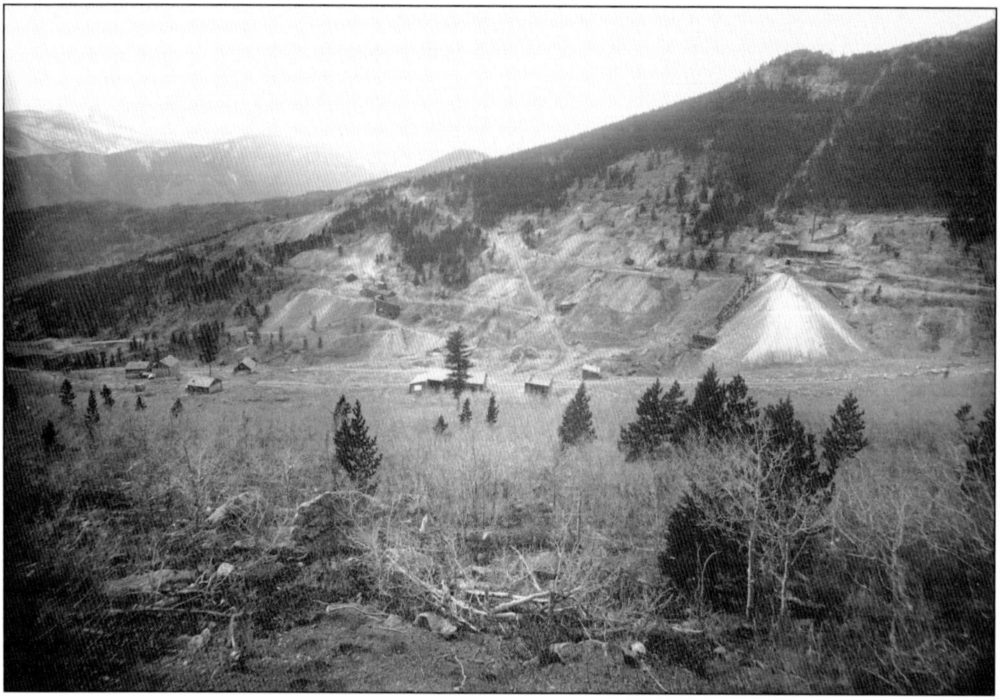

The mines of North Empire (Upper Empire) included Silver Mountain Mine (left), Tenth Legion (center), Gold Dirt Mine (right center), and the access road to the Benton Lode on Silver Mountain (center). John Dumont's Boston boardinghouse is visible at lower center. Dumont invested thousands around Clear Creek County. (UCB.)

GO TO EMPIRE!

Nowhere in the entire front range of the Rocky Mountains is there a town more beautifully, or more healthfully, situated than Empire, in Clear Creek County, Colorado, 45 miles west from Denver. A delightful place of residence all the year around. A charming summer resort. An important and rapidly developing gold mining camp, with lodes of true fissure character that may be worked generations to come.

Empire presents to the investor opportunities for assured wealth, and to the visitor a sojourn in one of the loveliest spots in the country. Short drives, or trips by rail, amid grand and picturesque mountain scenery, carry the tourist to points of world-famed interest, while home-like hotel accommodations provide material comfort during the few hours spent within doors in this land of sunshine. With streams teeming with trout for the angler; mountains teeming with the precious metals for the investors; grandeur of scenery for the lover of nature, and health-giving air and pure water for the invalid.

Empire extends a "glad hand" to all. Come and see us.

OVER

"Go to Empire!" reads the back of a postcard from the Colorado & Southern Railway in 1902 promoting the benefits of visiting Empire. This was a period when further local mine investment was sought to improve mines, build mills, and finance an industry. Money poured in from all over the world. Many Clear Creek County mining claims (and others statewide) are named for the wishes of foreign investment. Tourism also benefitted from establishment of the railroad. (CCHS.)

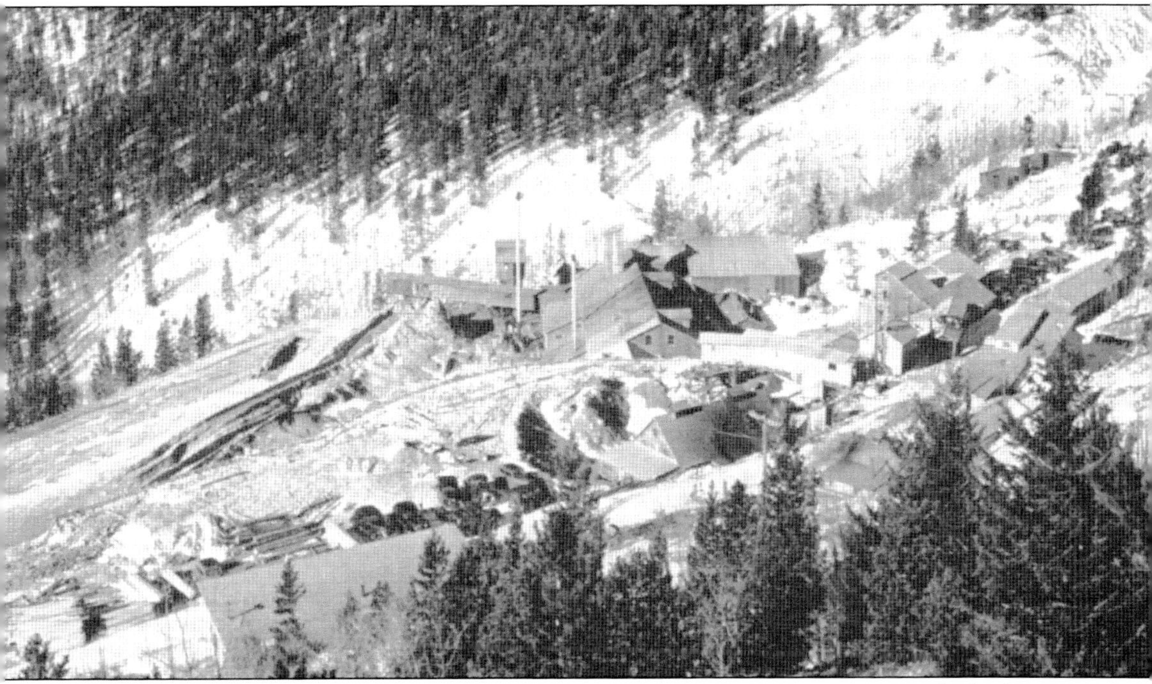

Seen here are the Minnesota Mines in 1940. Large, nationwide corporations were on the lookout for mining investments as the price of gold was recalculated at $35 an ounce in 1934 by FDR. The 3M Corporation and the Clifford family (owners) invested in the old Atlantic, Crown Prince, and Cleopatra claims in Lyon Gulch in North Empire. Lyon Creek is often incorrectly referred to as Lion Creek after the local mountain lions. The cyanide mill was constructed to process 250 tons of ore per day. (CCHS.)

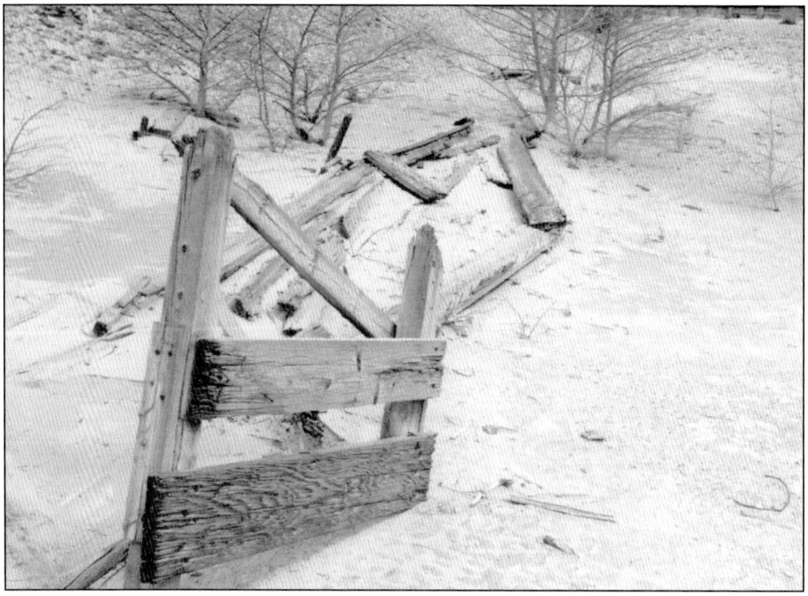

Remnants of the old flume in the tailings at Empire are seen here in 1996. No traces of these flumes remain today. The flume ran all the way down Lyon Creek from North Empire. (DPL.)

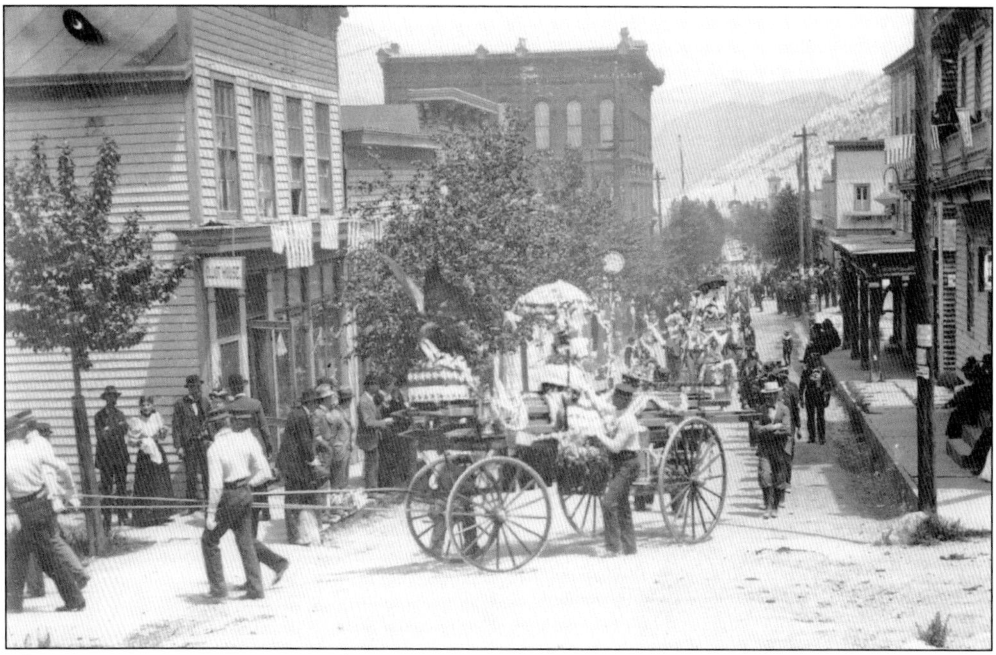

The Empire Hose Cart is on parade in Georgetown in 1897. With the Elliott House on the left and the Ennis House on the right, the hotels in Georgetown were usually full of potential investors, who would travel from distances far away to see about investing in the gold and silver mines of the area. Indian fighter and Civil War hero George A. Custer invested in Georgetown silver mines before meeting his demise at the Battle of the Little Bighorn. Custer's supporting military commanders still carried resentment over Custer's mishandled raid at Washita, Oklahoma, in 1868 and did not come in time when they were summoned in Montana.

The tailings of the old Minnesota Mines looked like the craters of the moon. There was so much toxic waste from the gold-reducing cyanide at the North Empire location as well as the tailings in lower Empire. There are tales of children playing in the mine tailings in the early days when the toxic waste was fresh and then all getting sick with a variety of diseases caused by cyanide. (Lynn Schimmelfeder.)

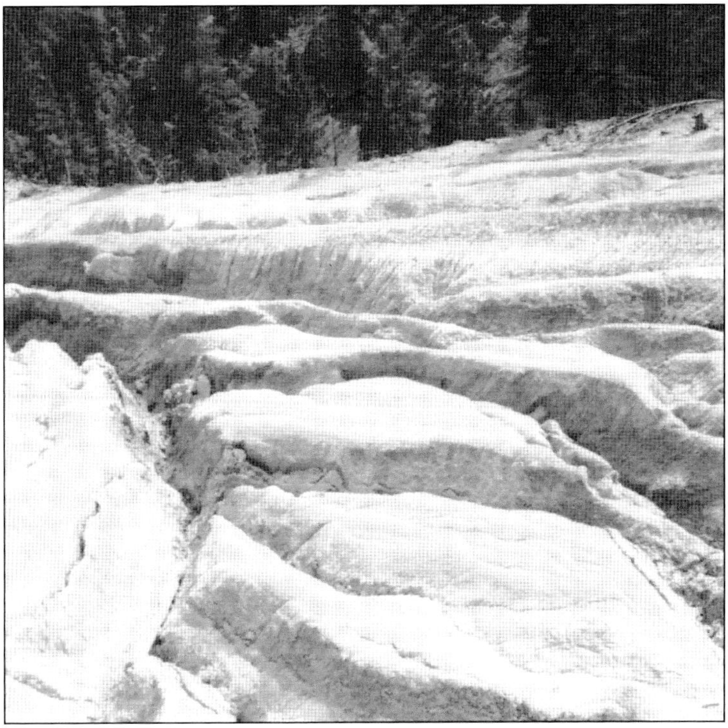

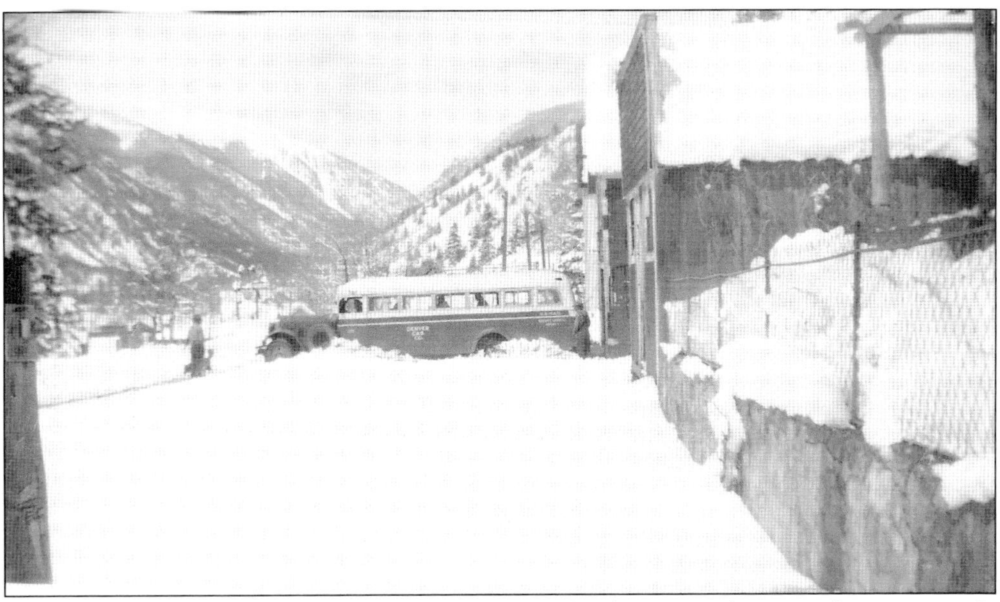

This view of Empire, looking west in the 1950s, shows a bus being stored in Maire barn, previously located in a grass field in front of the Peck House. The Maires lived in a house next to the pioneer postmaster Avery's home. Neither the barn nor the Maire house remains today. The Maires ran a hack service from the train depot to Empire.

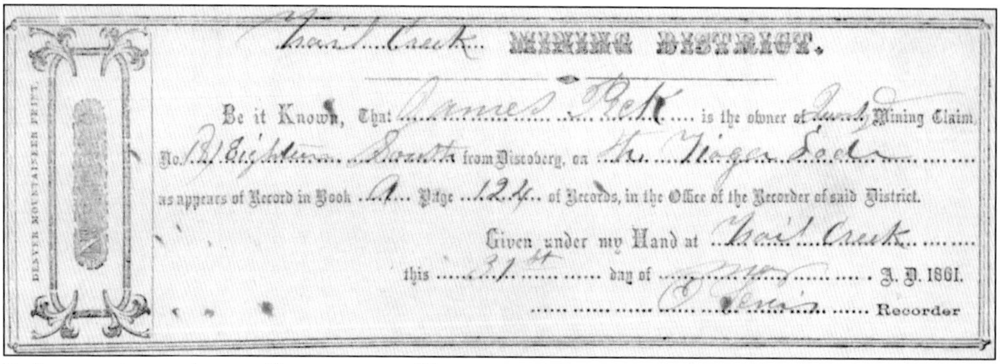

Trail Creek MINING DISTRICT.

Be it Known, That James Peck .... is the owner of .... Mining Claim No. 18 Eighteen South from Discovery, on the Tiger Lode .... as appears of Record in Book A Page 124 of Records, in the Office of the Recorder of said District.

Given under my Hand at Trail Creek ....

this 31st .... day of .... A. D. 1861.

.... Recorder

DENVER MOUNTAINEER PRINT.

Before coming to the pro-Union town of Empire in 1861, James Peck tried his hand in the mining district of Trail Creek near Idaho Springs. During the Civil War, people, as well as towns, were either Yankees or Rebels. There were too many rebels in Trail Creek for Peck's liking, so he moved on to Nevadaville and managed the Whitcomb Mill. He then heard about the benefits of a new mining camp at Empire after miners from Nevadaville crossed a few mountains and discovered silver and gold at Silver Mountain in North Empire. (DPL.)

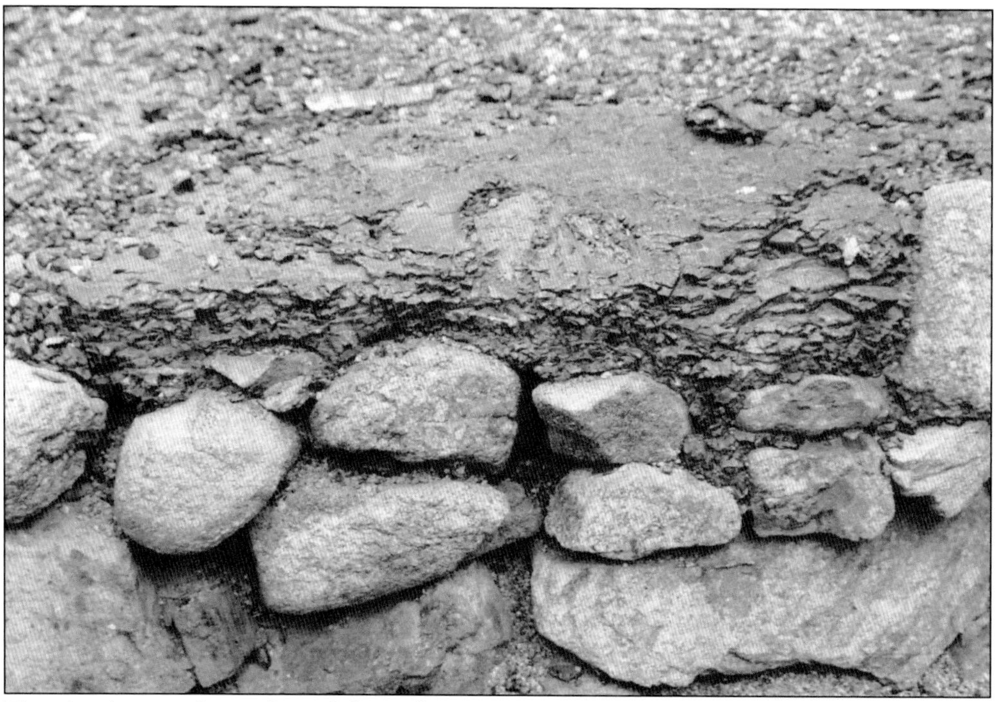

This slag (a waste by-product of the milling process) is from the Denver City Mill, which was later refurbished as the Kistler Mill.

Five

Dumont, Lawson, and Silver Creek

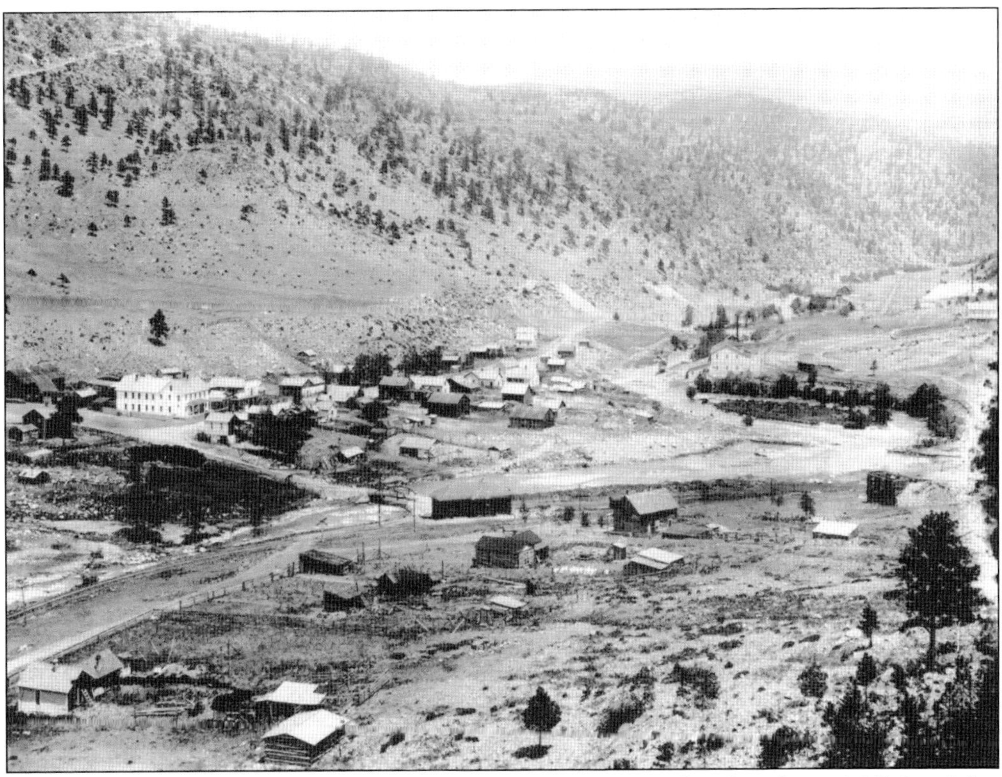

The community of Dumont, seen here around 1880, was previously referred to as Mill City. John C. Dumont poured so much money in the area that the name was changed to honor him. He was a mining speculator and investor. He bought mines at Hukill Gulch near Idaho Springs and across Clear Creek from the Stanley Mines, and only then did they become profitable. He invested in mines and smelters in Georgetown and North Empire. The Mill City house at left center in this photograph was one of the finer hotels in the mountains. The railroad depot is situated across the Clear Creek at lower center, and the train grade passes directly behind the mill located on the creek (right center). (DPL.)

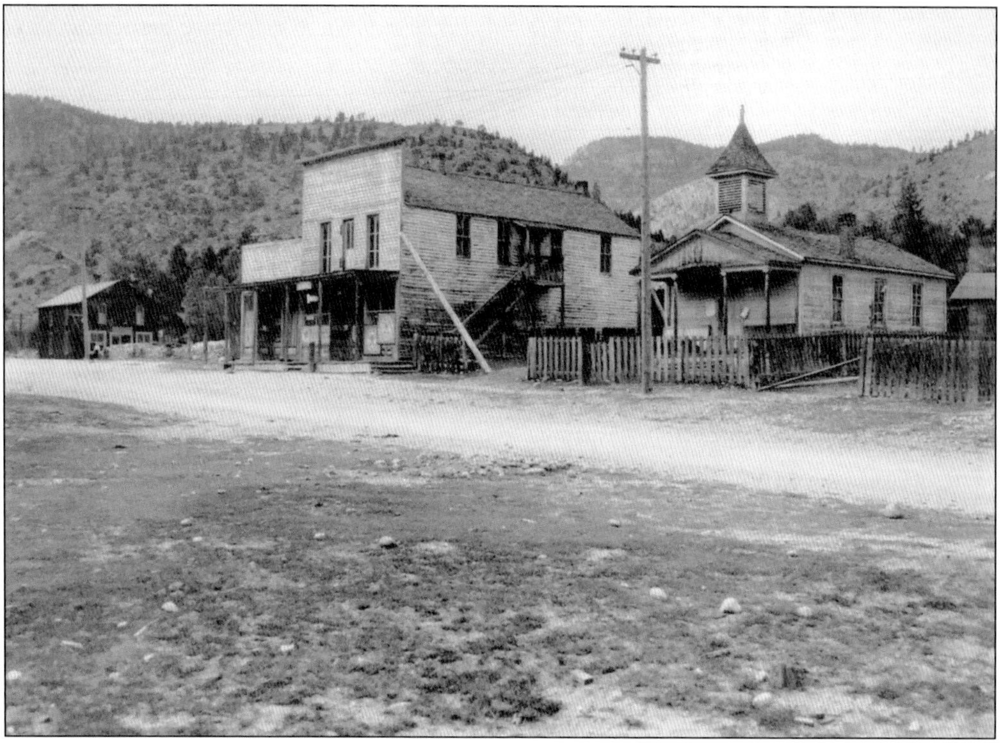

A mercantile and small church were located at Dumont. None of these structures remain today. (DPL.)

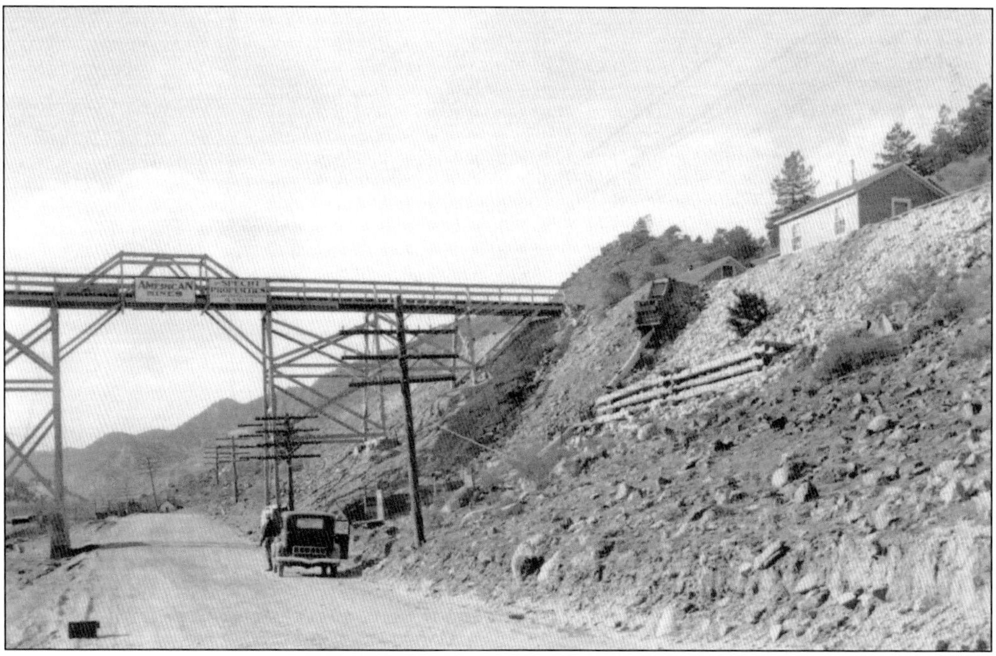

The American Mines and Specht properties operated the Clear Creek and Gilpin Tunnel at Dumont. It was designed to penetrate the mines at Russell Gulch in Gilpin County. Ore was carted across the trestle over unpaved Route 40 to the mill across Clear Creek. (DPL.)

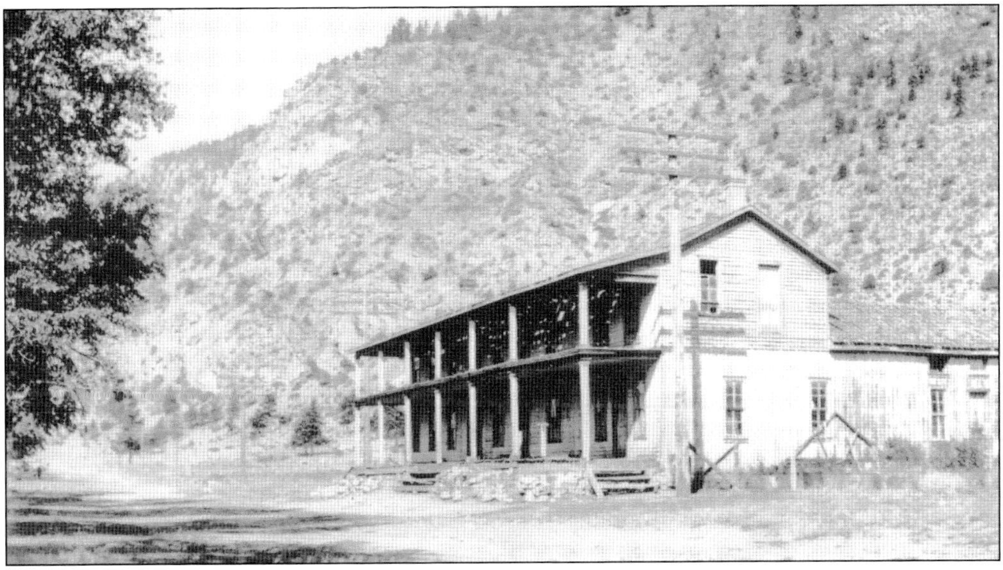

Alex Lawson, who courted the daughter of John Coburn, who came to the area in 1871 and founded Free America (later Downieville), built the six-mile house at Lawson. By the time this photograph was taken, one can visibly see the holes in its roof. (DPL.)

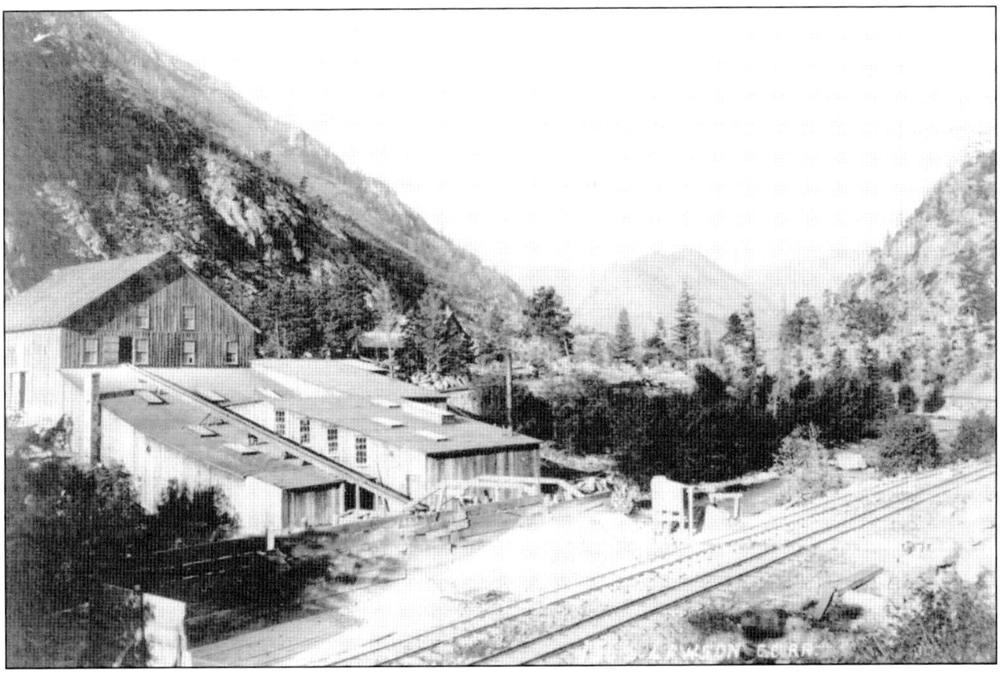

The Joe Reynolds Mill at Lawson, built in 1900 along the Colorado & Southern right-of-way, served the mines at nearby Silver Creek and Red Elephant. The miners at the company town of Red Elephant became so disgusted with their mine bosses that they all quit and went to work across the valley at Silver Creek. Soon, after the price of silver was forced down, they all lost their jobs. History often reveals blunders of bad timing. (History Colorado.)

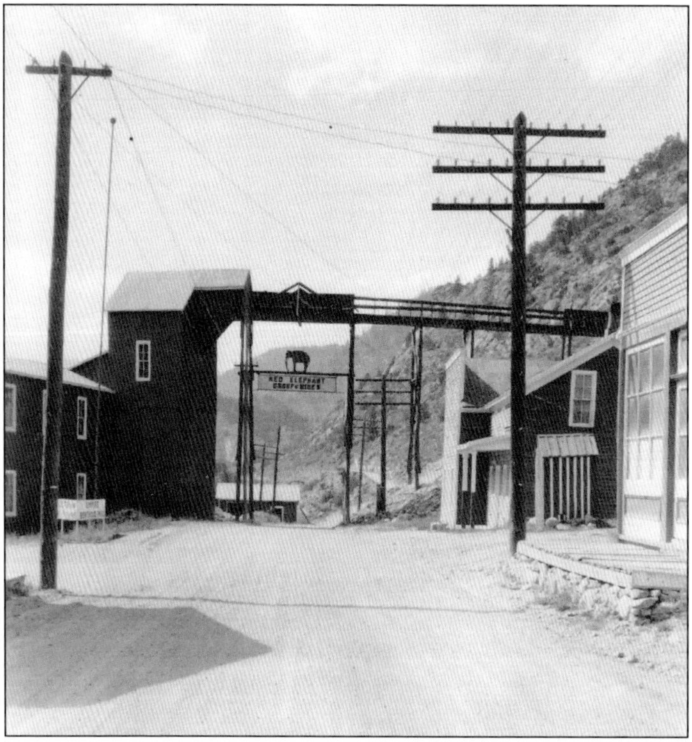

The trestle over Route 40 took the ore from the mines at Red Elephant to the mill along the creek. Each time a mining claim was awarded, a mill site was also assigned along the creek. In some cases, this was practical, and in other cases, it was more practical to power the mill with a large boiler. (DPL.)

The McClelland Mine sits at lower left with the large dump of tailings, and the Clear Creek Gilpin Tunnel workings are at left center. The US Bureau of Public Roads improved the Route 40 along the creek in the early 1920s. (UCB.)

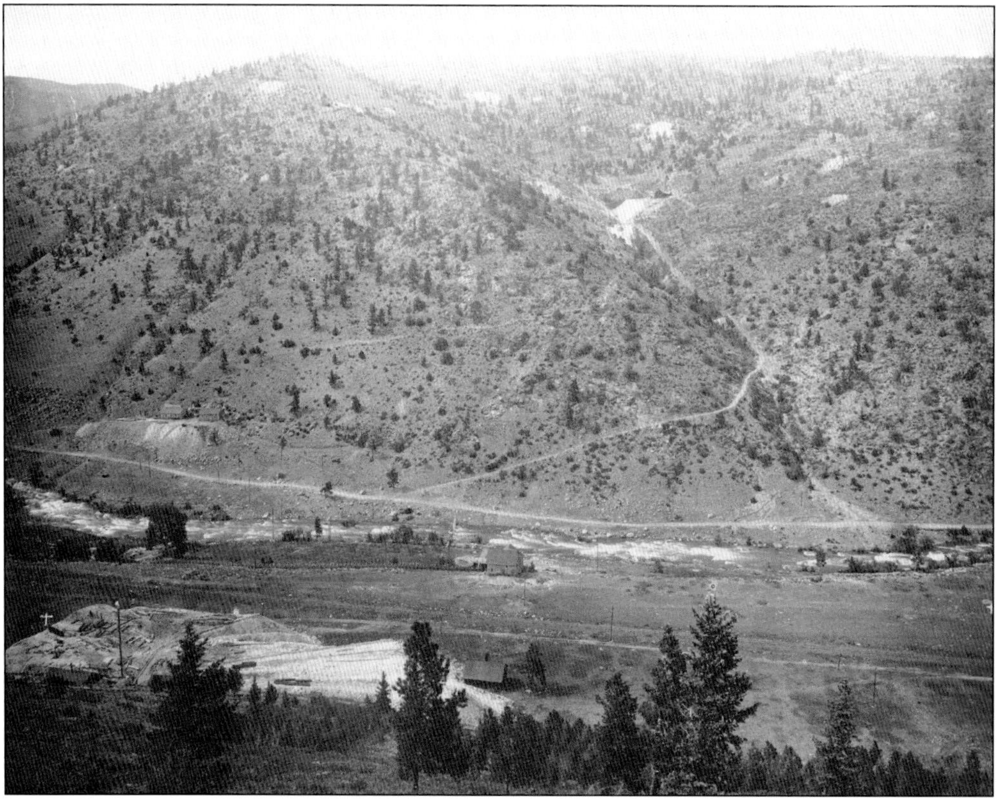

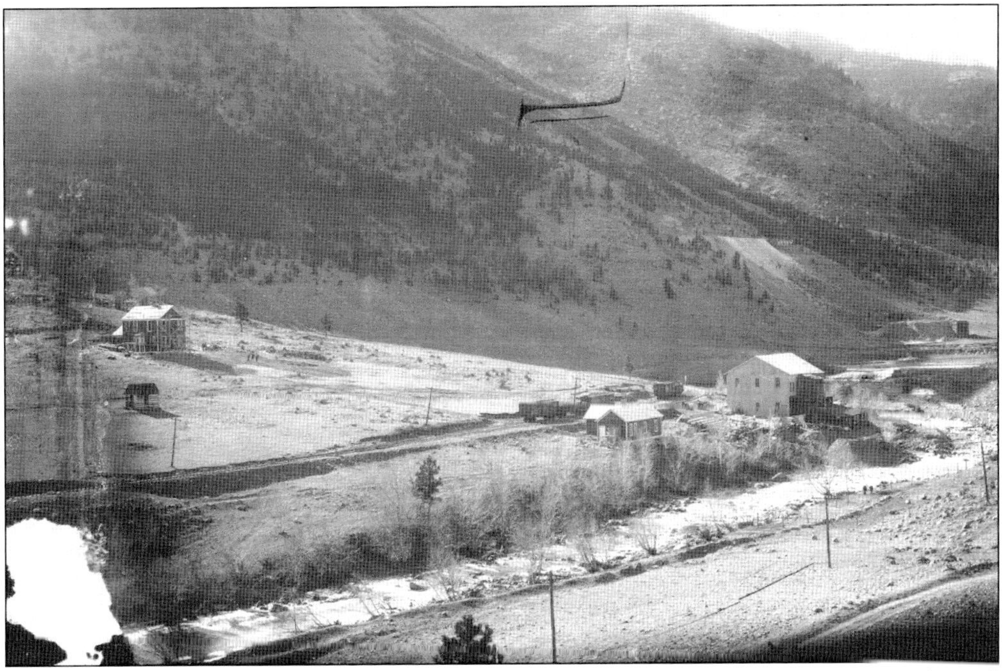

Also pictured on page 103, this mill is shown from the opposite side; the railroad access to the mill is clearly visible. In the 1930s, when the Clear Creek Gilpin Tunnel was reopened, a trestle was built across Clear Creek. Note the house under construction at left center. (History Colorado.)

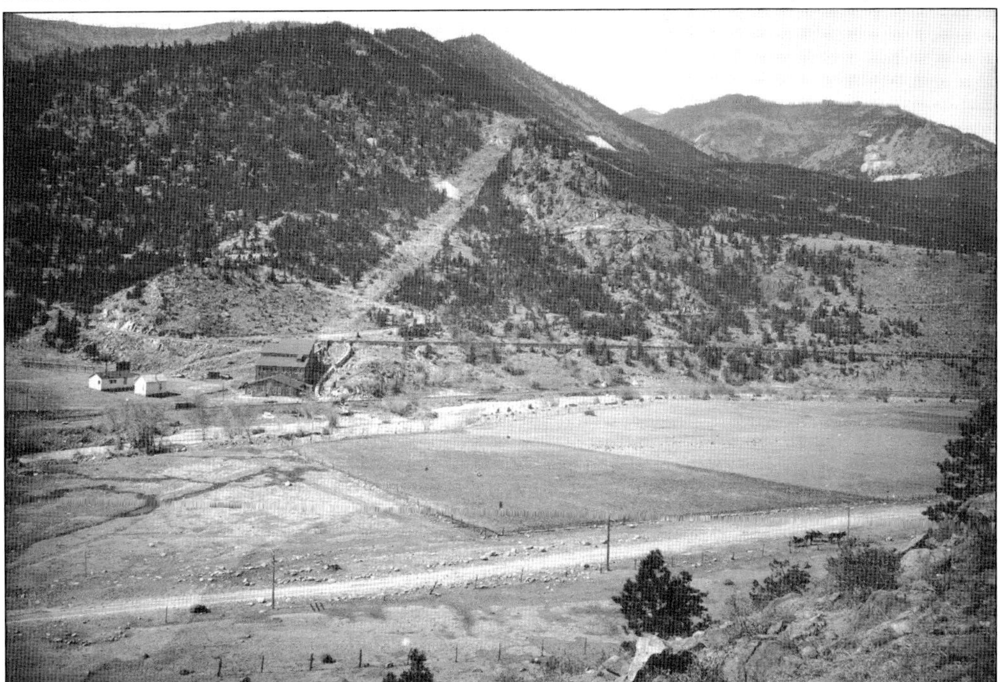

Here is a view of Downieville with the mill across Clear Creek near the Colorado & Southern rails and the wagon road to Georgetown in lower front. Note the American Sisters Mines at Silver Creek in the upper right of the photograph. (UCB.)

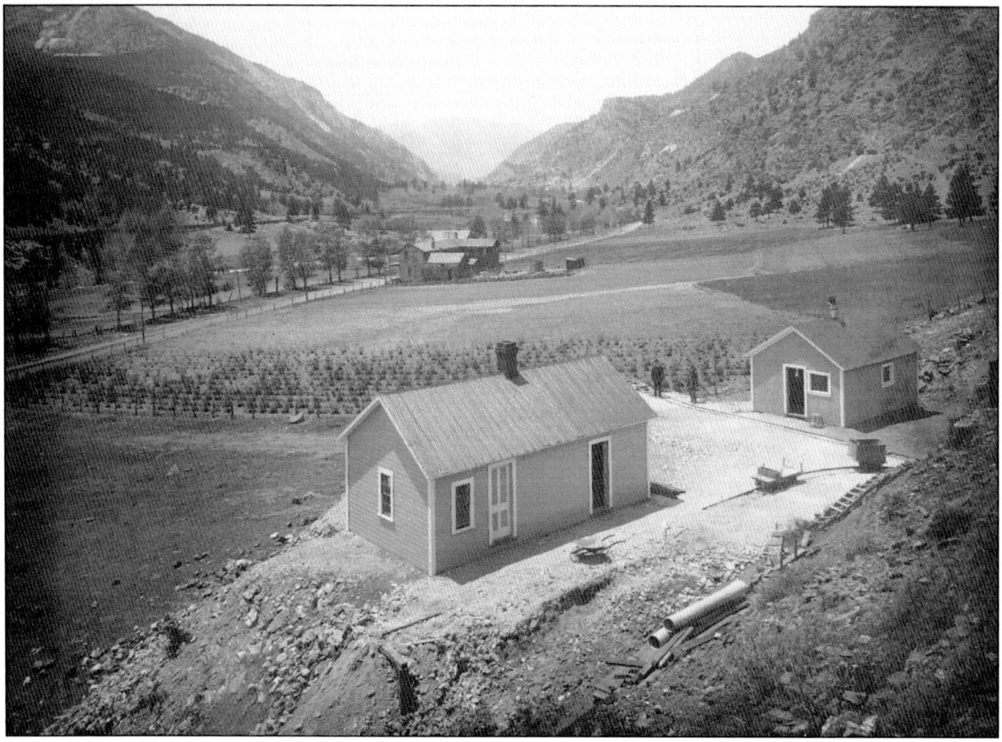

The 4C Mine at Downieville is pictured along with the old six-mile house in the center. This area of the valley had been used for farmland. To this day, the mine dump tailings can still be seen behind the Conoco and Starbucks in Downieville. (UCB.)

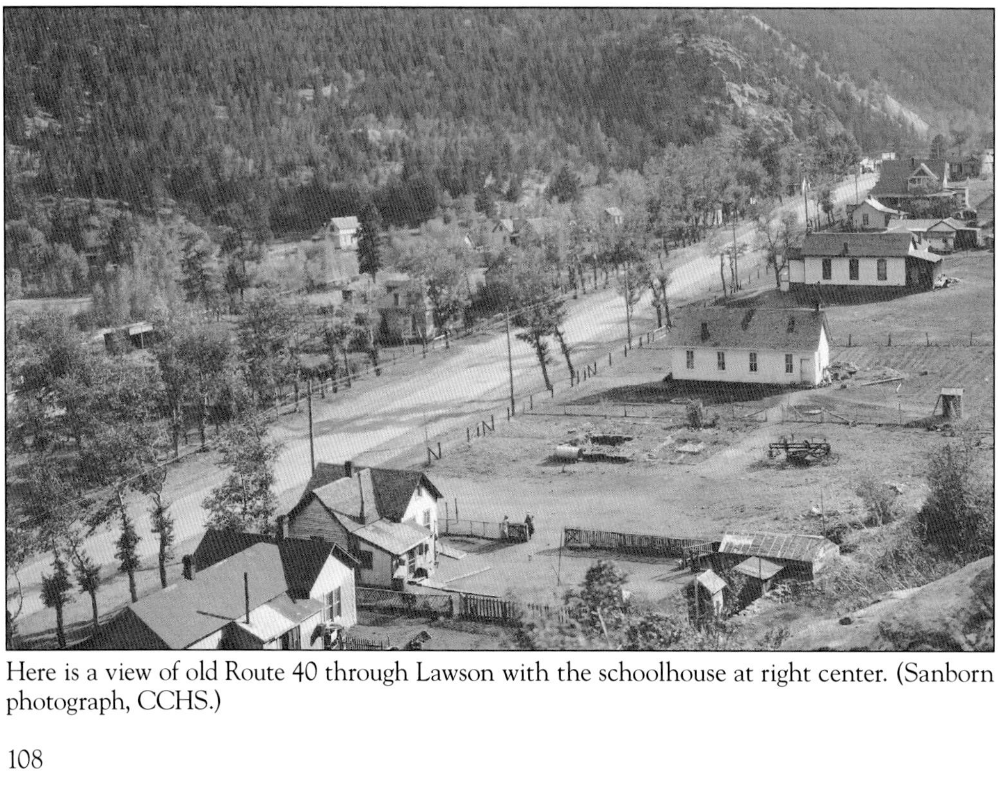

Here is a view of old Route 40 through Lawson with the schoolhouse at right center. (Sanborn photograph, CCHS.)

Six

CLIMAX, URAD, AND HENDERSON MINE

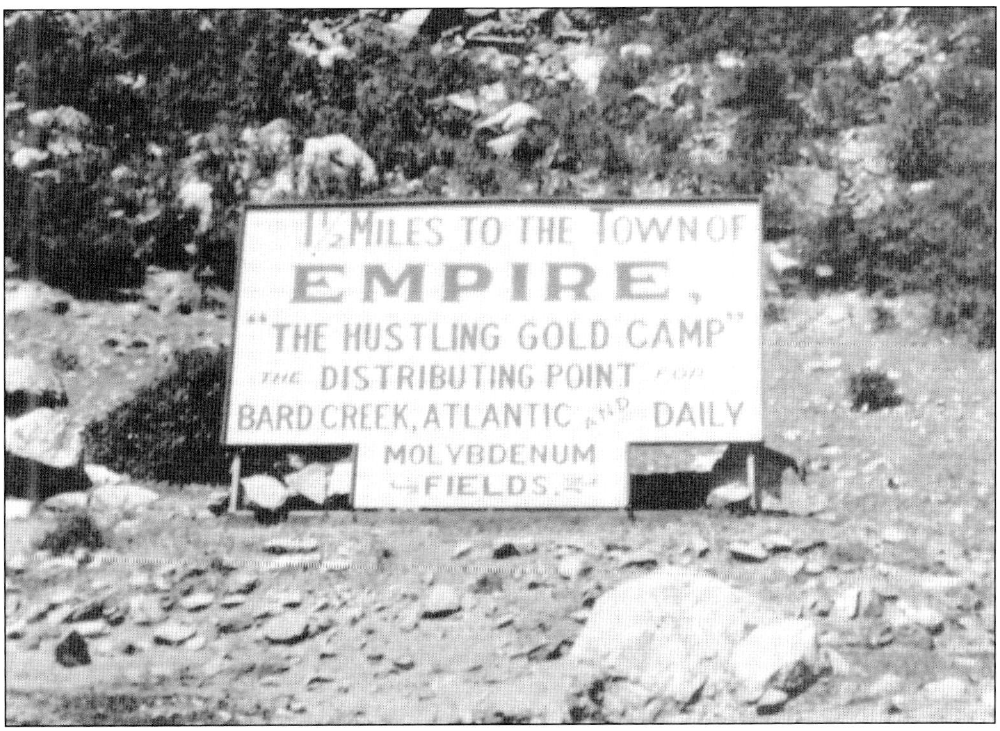

An early roadside sign from 1920s advertises the mineral riches of Empire as well as those of Bard Creek (south of Empire toward Georgetown) and the Atlantic and Dailey district's molybdenum fields, which became the URAD (Uranium Research and Development Corporation) and Henderson Mines west of Empire.

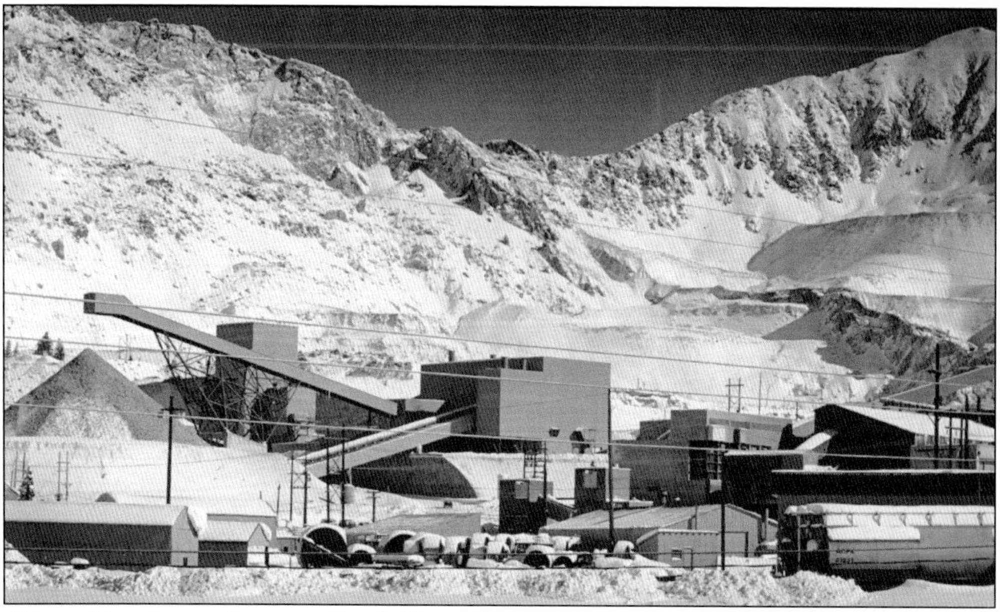

Climax Mine is seen here near Leadville in 1982. Climax opened in 1879 as a silver mine and later mined an important steel additive, molybdenum. Charles J. Senter discovered the claim in 1879 and held onto it for 39 years. Climax's official startup year was 1918, for the military. The properties of molybdenum were used in manufacturing arms for World War I. Early uses of molybdenum were unknown until C.H. Mills, a metallurgist working for Henry Ford who designed the Ford logo, used molybdenum to strengthen his car. Then, vast markets for the mineral were found. Climax development continued through the 1980s, and it still operates to this day. (*Denver Post* photograph, CCHS.)

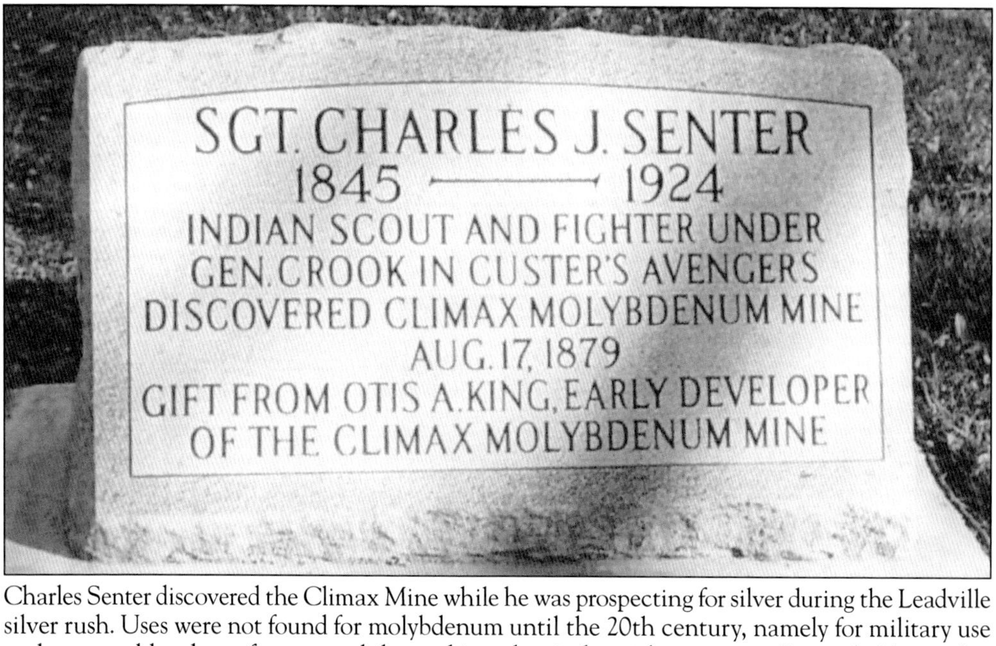

Charles Senter discovered the Climax Mine while he was prospecting for silver during the Leadville silver rush. Uses were not found for molybdenum until the 20th century, namely for military use and as a steel hardener for automobiles and in other industrial capacities. Senter held onto his claim for 39 years, until he was paid $40,000 in 1918. He retired in comfort in Denver.

James Ludwig was named superintendent of the Climax Mine in 1965, and the mine was developed under his watch. In 1967, the mine began a period of drilling and expansion.

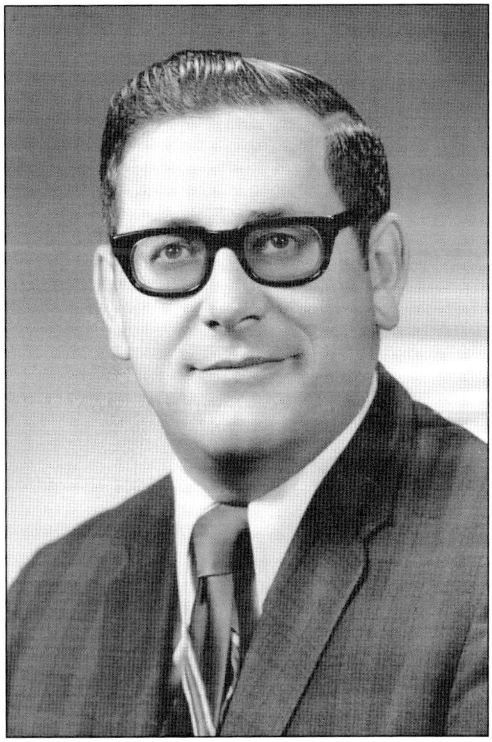

This 1926 advertisement is for the Wills Sainte Claire, the first automobile to utilize molybdenum in the manufacturing process. Metallurgist C.H. Mills left Ford motor company and designed the automobile around a World War I aircraft engine.

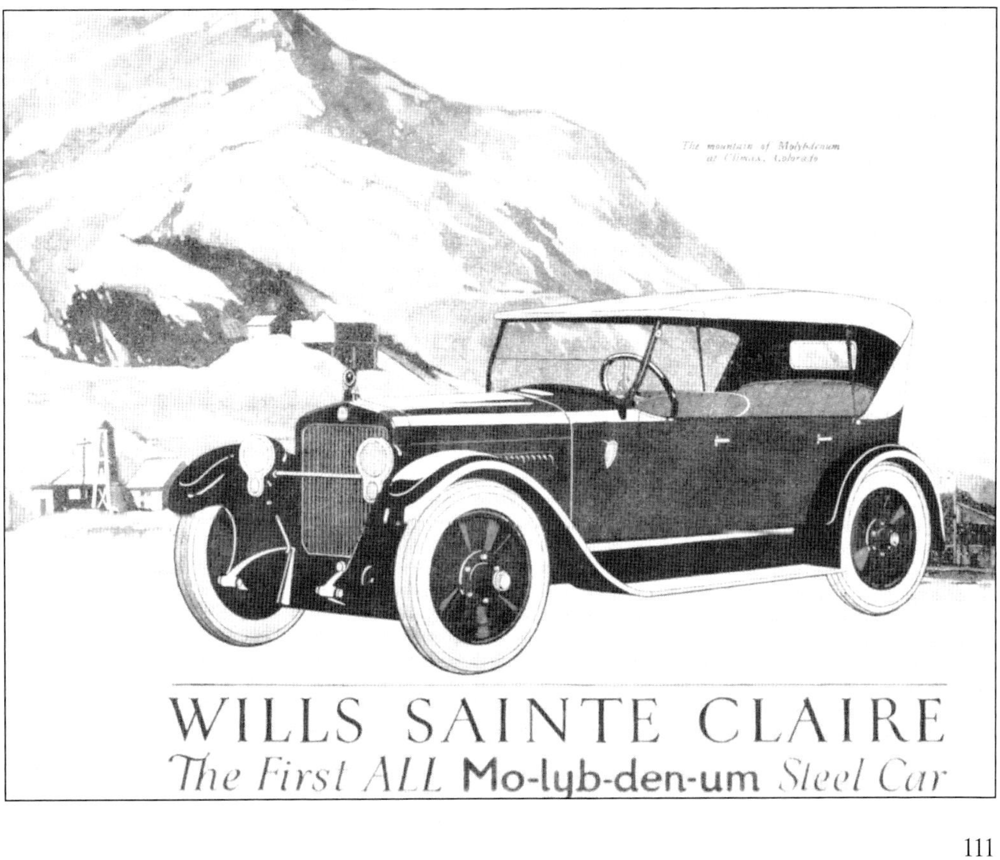

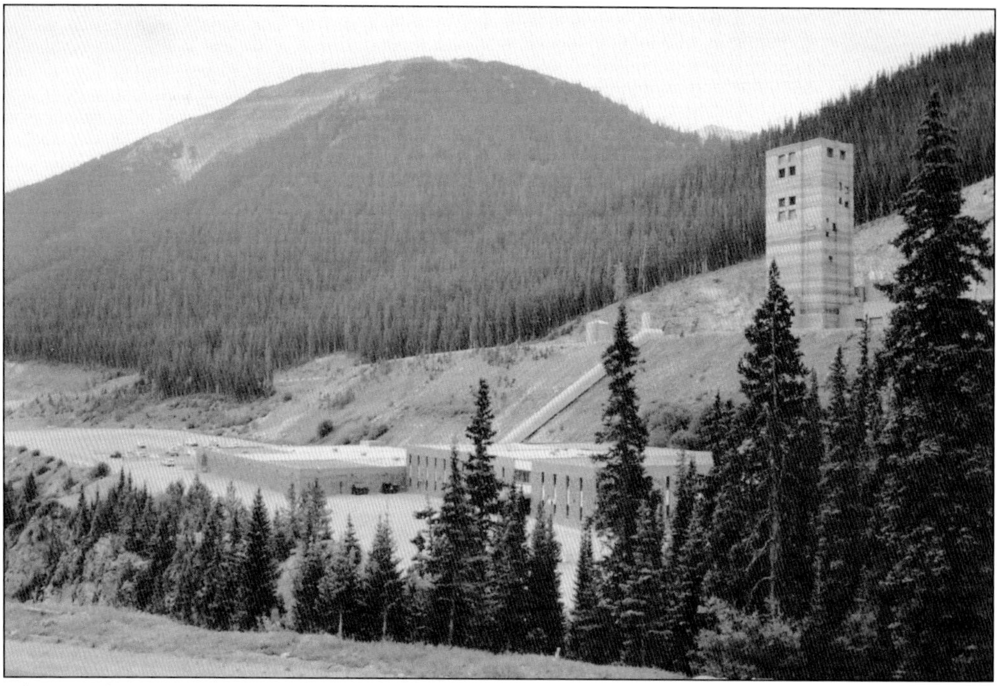
The Henderson Mine was developed in the late 1960s and early 1970s. Millions of dollars of investment was required by Climax/Amax to get the mine properly developed and to build the mill 13 miles under the Continental Divide.

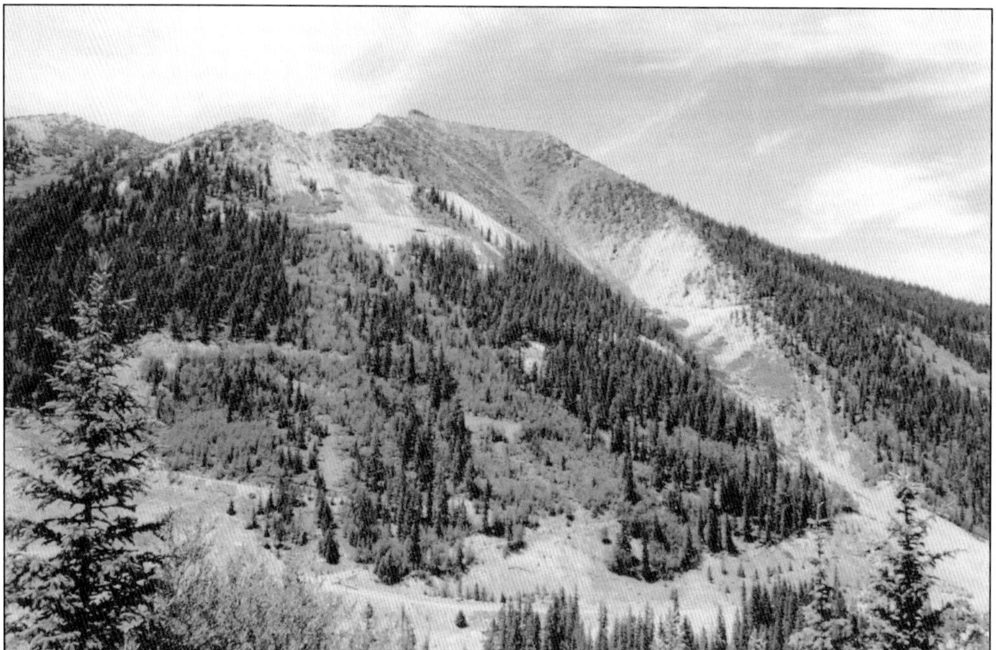
Red Mountain is where millions of tons of molybdenum lay buried until Climax authorized test drilling in the 1960s. Initially, the URAD Mine was developed on the south side of the mountain until further drilling was done on the other side of the mountain in 1964.

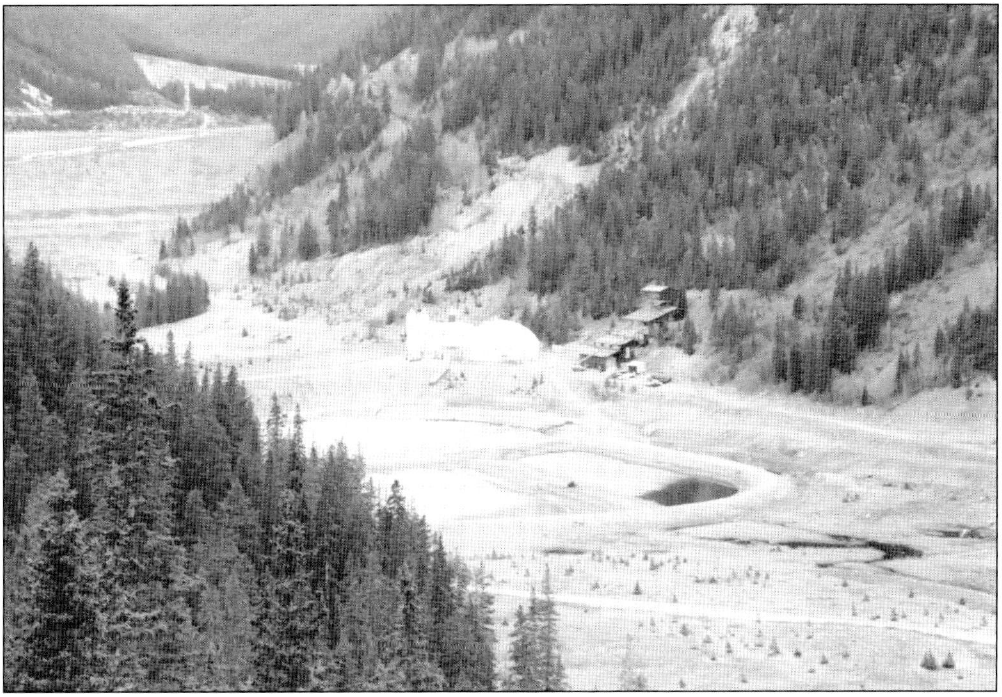

This valley was the site of the URAD Mine, and thousands of tons of ore were mined out before the site was reclaimed in the 1970s. Climax/Amax received several environmental awards as a result of its reclamation efforts.

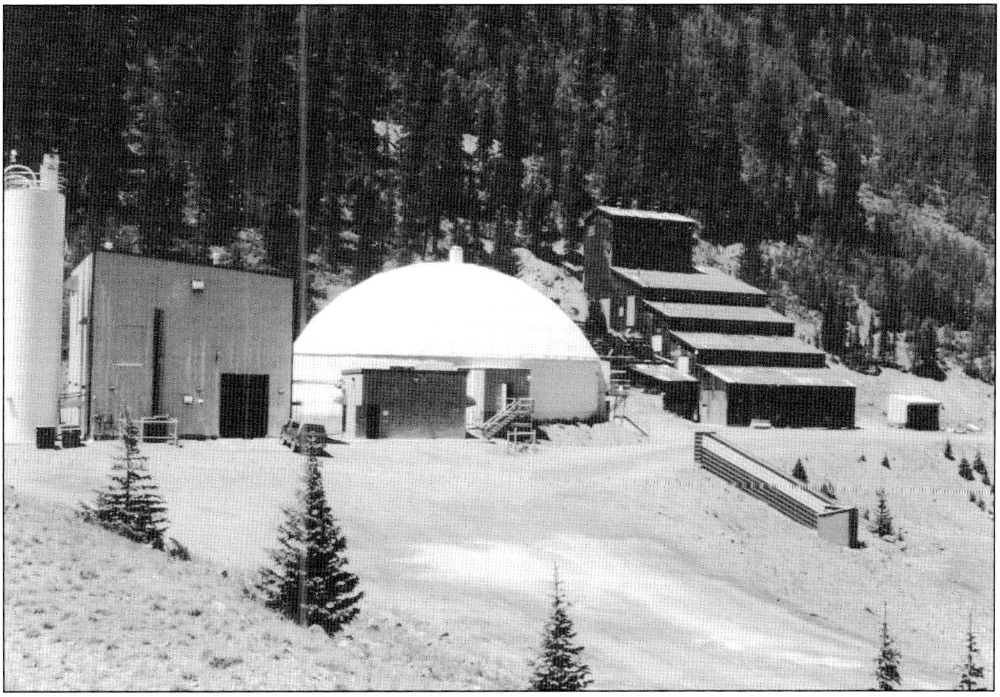

Pictured here is the mill and water treatment plant at URAD. The equipment under the dome processes 100,000 gallons of water daily.

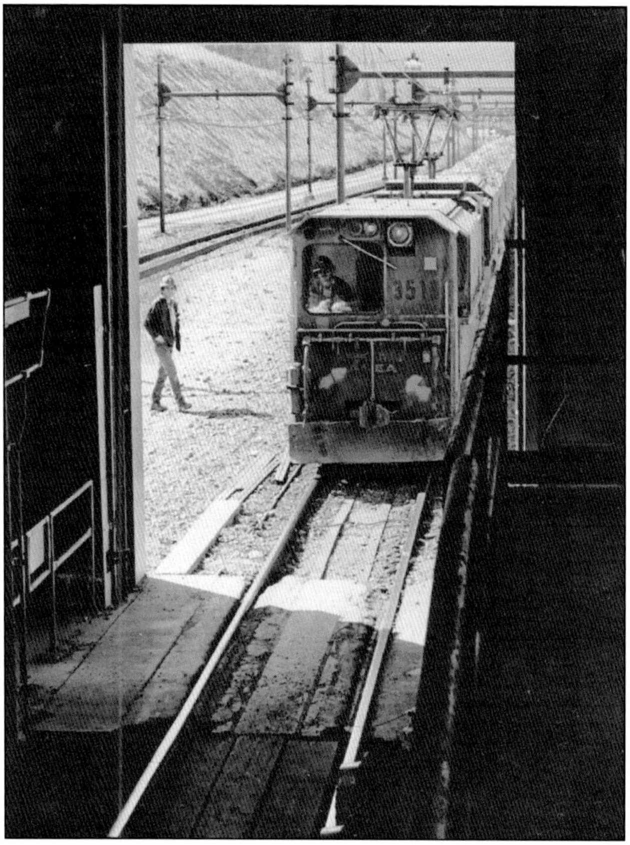

In this 1979 photograph are the old railcars used to haul the molybdenum ore from Henderson Mine near Empire underneath the Continental Divide to the mill near Parshall.

URAD Lake in the URAD Valley, west of Empire, is pictured in this photograph. Berthoud Pass and US 40 wind up the mountain range in the background.

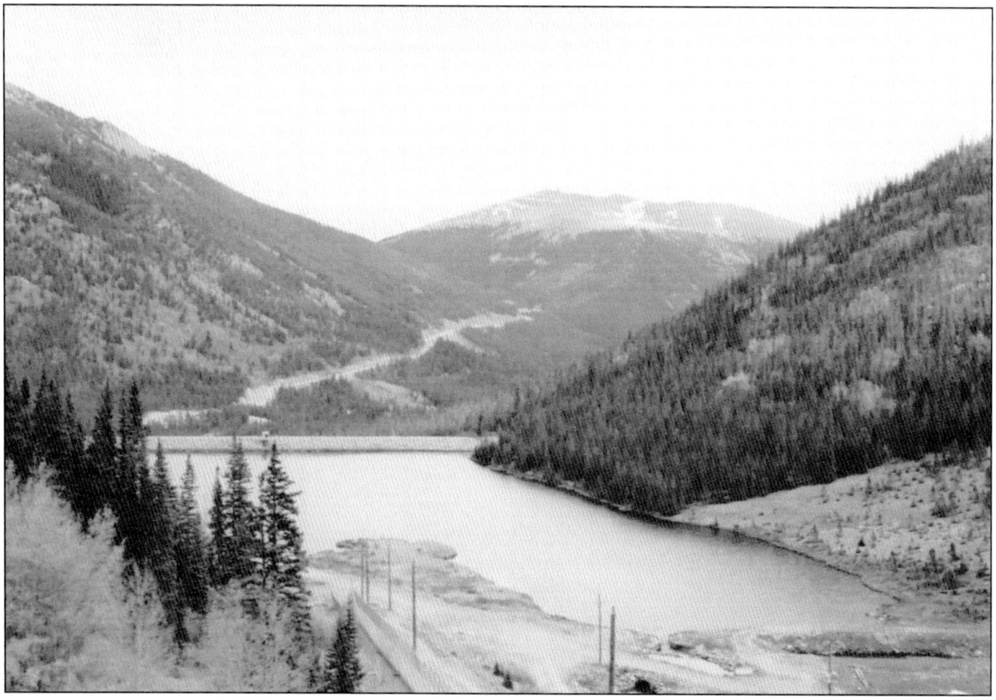

The Climax Molybdenum Co. placed this ad in 1937. Climax advertised that having a solid background on the benefits of using molybdenum in one's manufacturing process would be beneficial to a company.

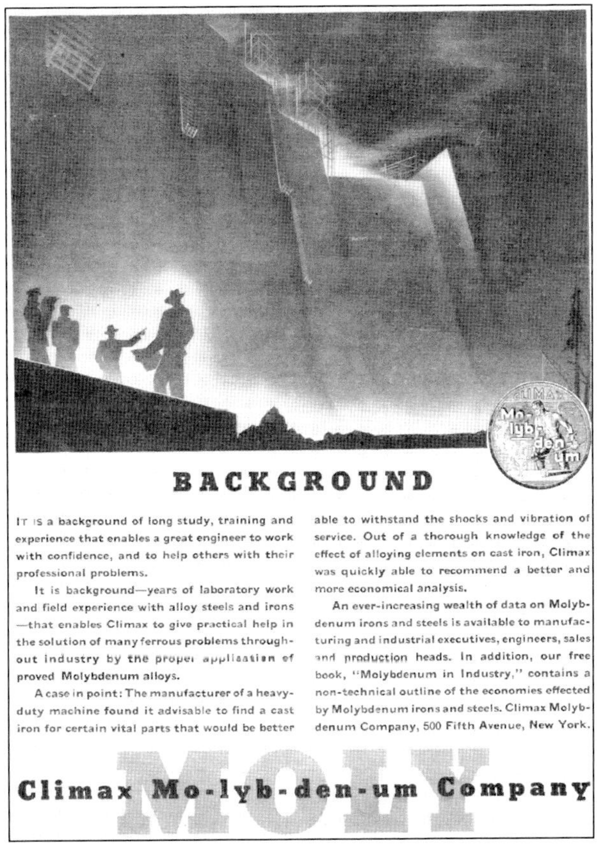

This photograph captures work crews building the crusher at Henderson in 1967. While Henderson was being built, an underground tunnel was also under development to carry the ore underneath the Continental Divide to the mill at Parshall. (DPL.)

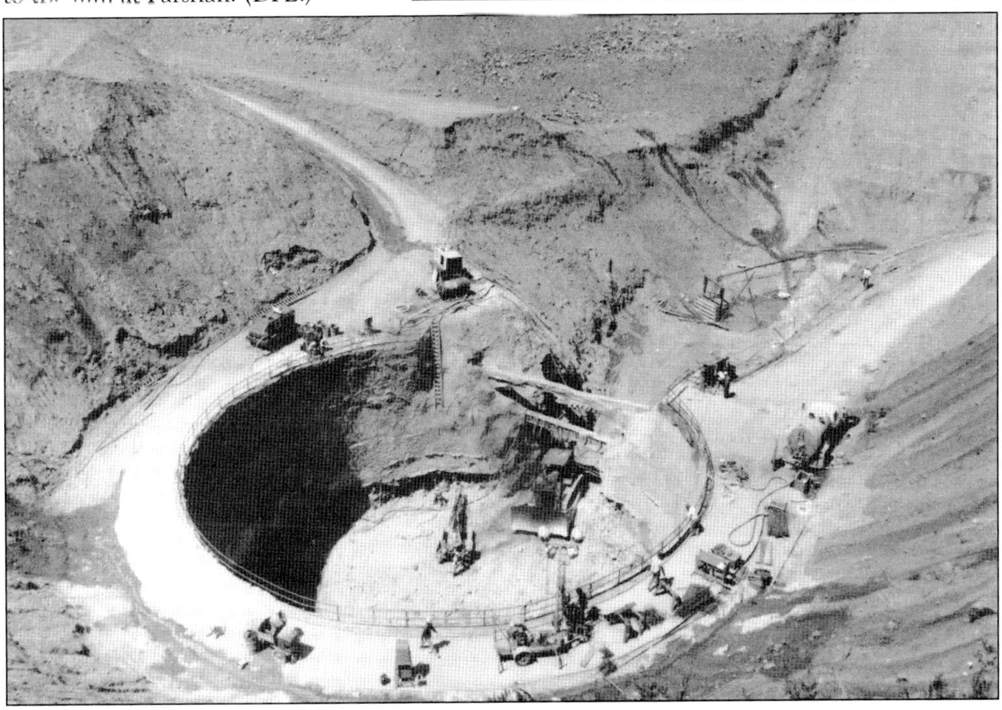

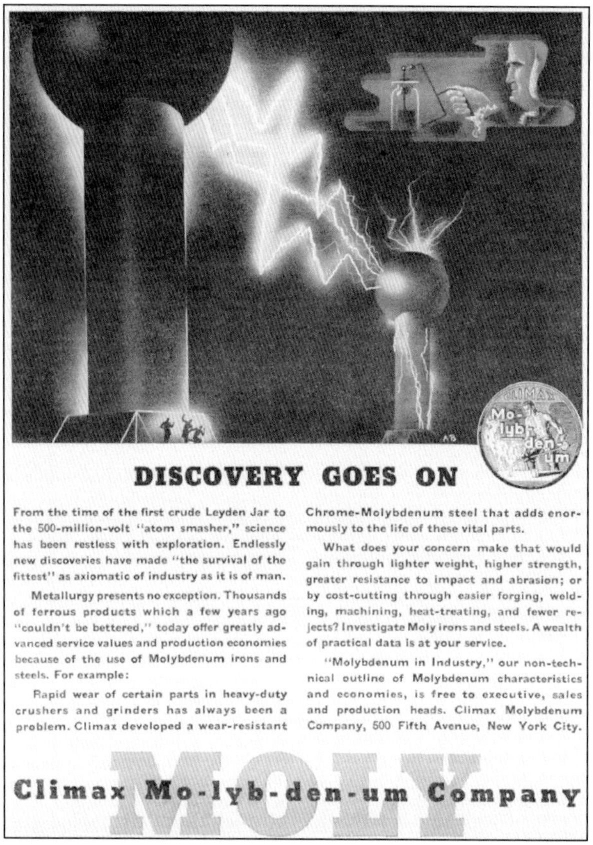

This Climax "Discovery Goes On" advertisement from 1948 explains that man is constantly making new discoveries and how a company would benefit from using molybdenum in its manufacturing process.

A miner puts lids on the 55-gallon drums so that the molybdenum can be shipped to the smelter in Iowa.

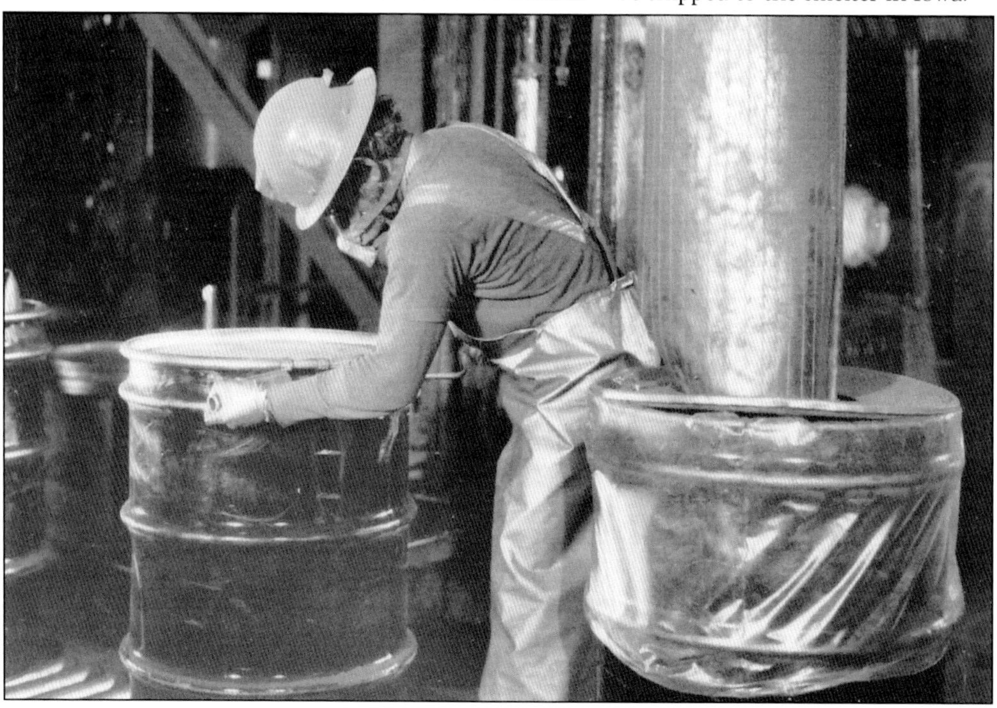

Here is a 1949 ad from Climax Molybdenum Co. in New York, New York, touting the advantages of using molybdenum in one's manufacturing process.

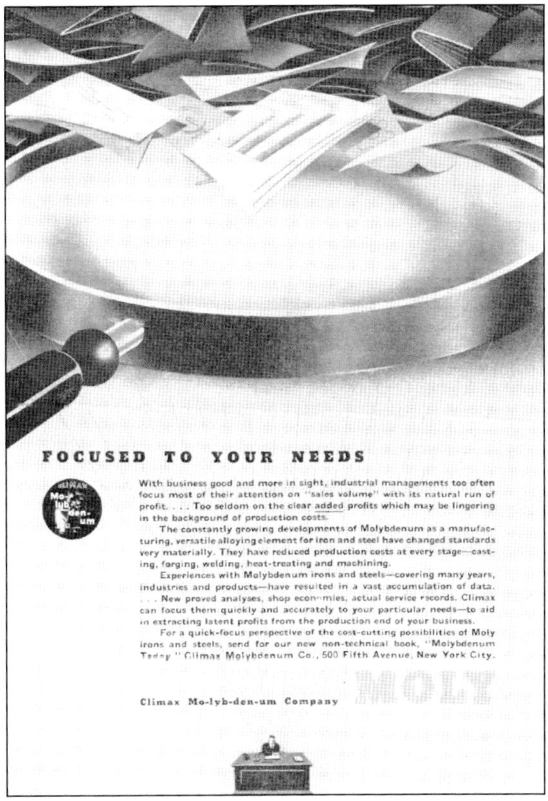

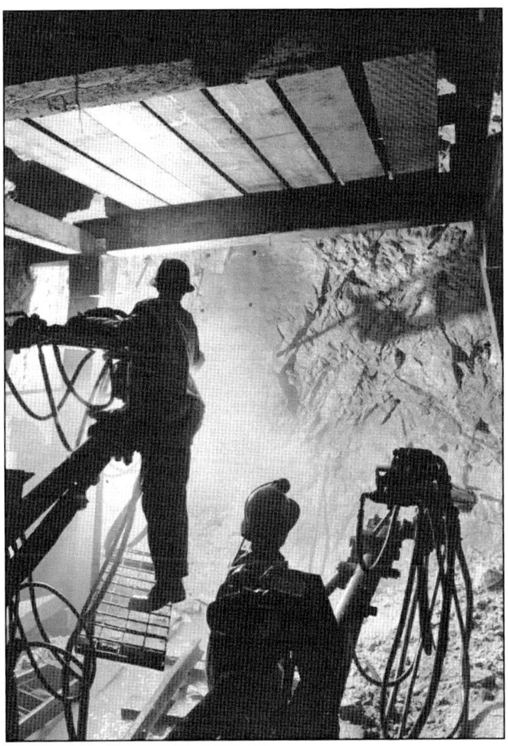

Workers blast a labyrinth of tunnels at Climax in 1967. Air drills are being used, and there is plenty of water on the ground.

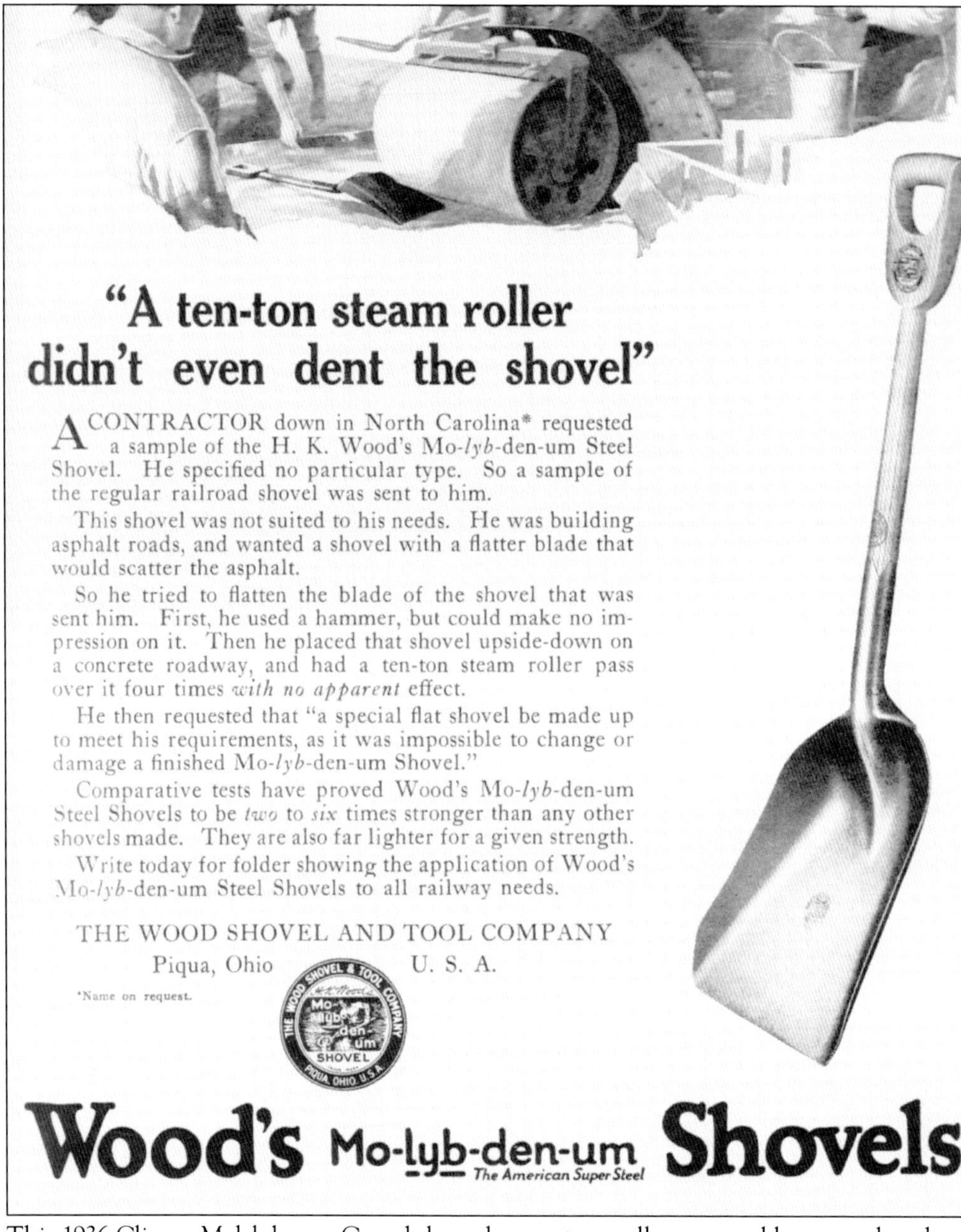

This 1936 Climax Molybdenum Co. ad shows how a steamroller was unable to smash a shovel that had been made with molybdenum.

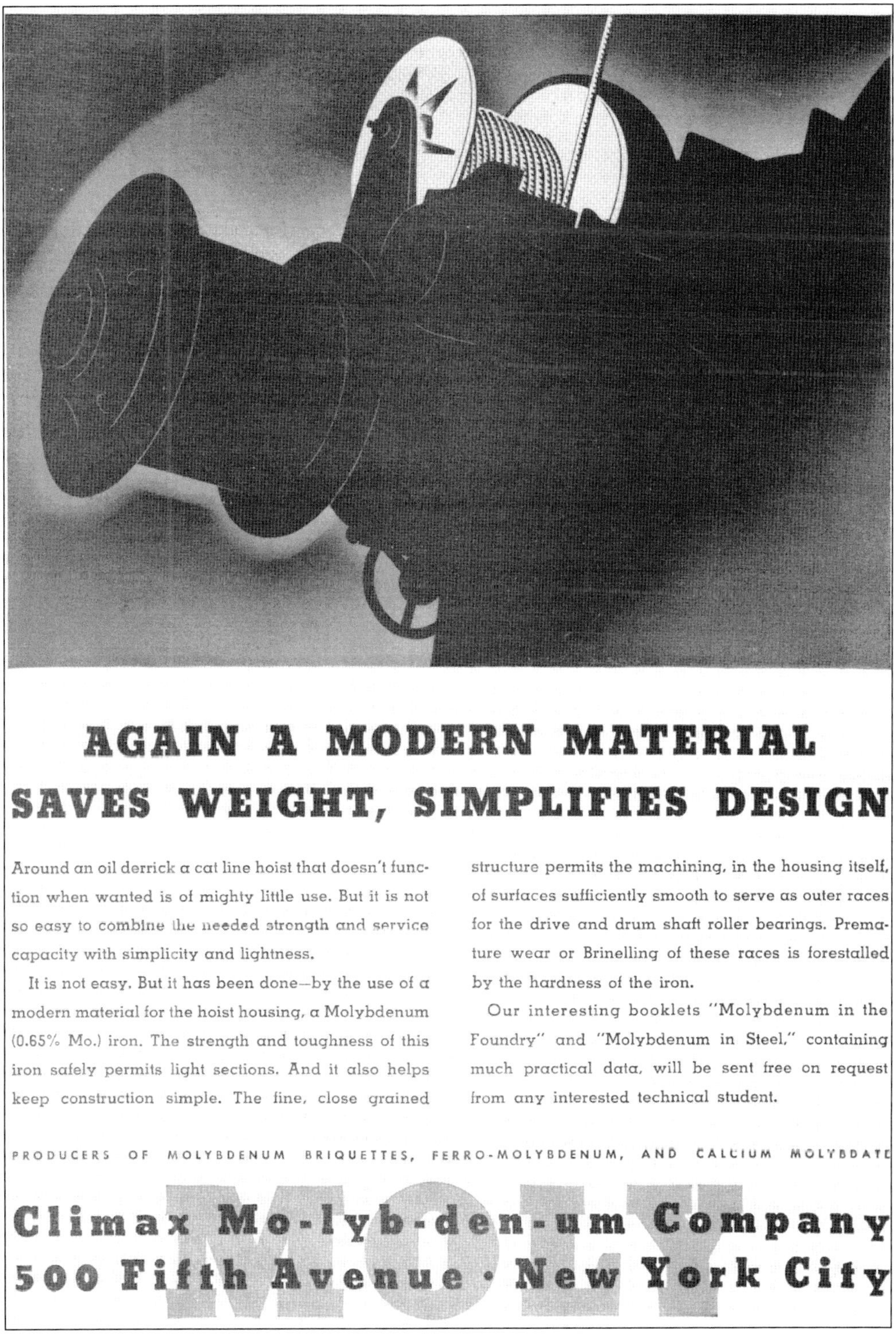

This is a 1953 ad from Climax Molybdenum Co.

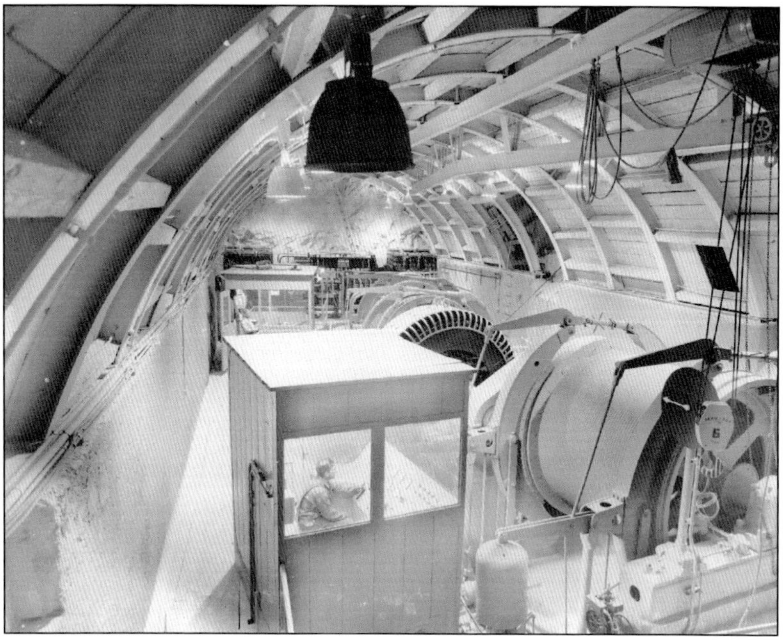

Pictured here is the control booth at Climax Mine near Leadville.

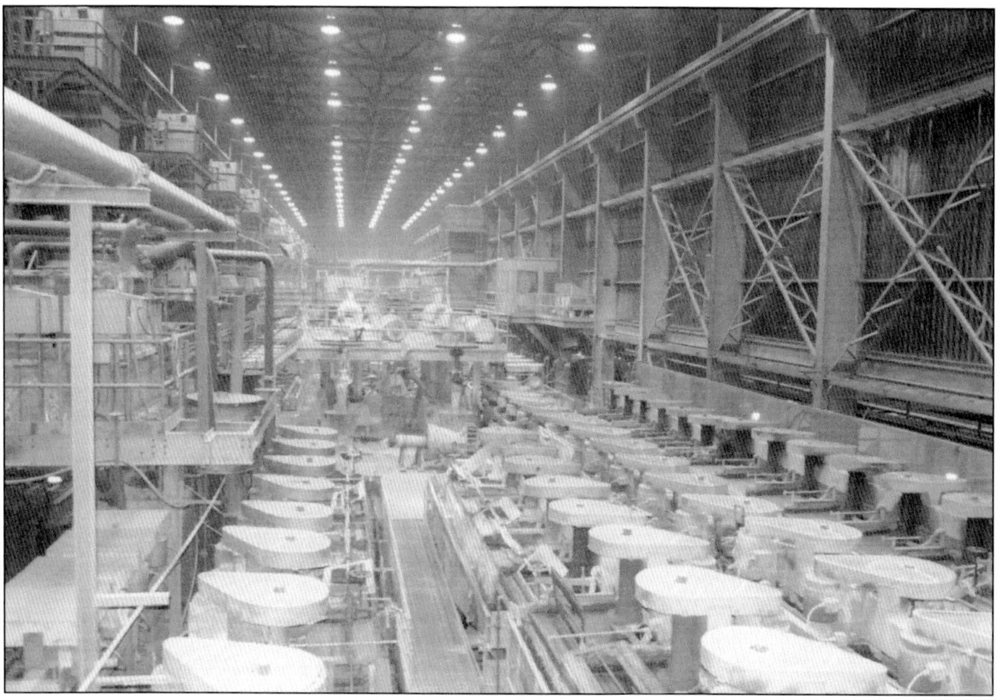

This interior view of Henderson Mill shows rows of flotation cells. Flotation cells are the last part of the milling process. The powder concentrate is mixed with water and separated after floating in the water for a period of time.

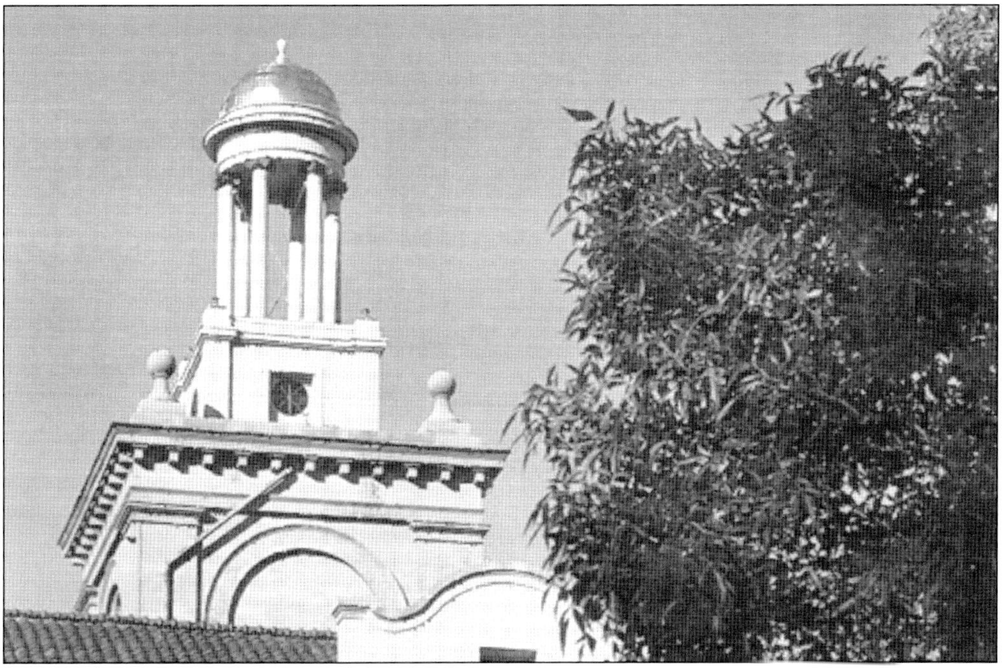

Pictured here is Guggenheim Hall at the Colorado School of Mines. Colorado School of Mines was and continues to be world renowned in advanced engineering degree programs, including mining engineering, petroleum engineering, and environmental engineering. Colorado School of Mines has its roots in the Colorado gold rush and began operating in 1874, first as a private school and then as a state school.

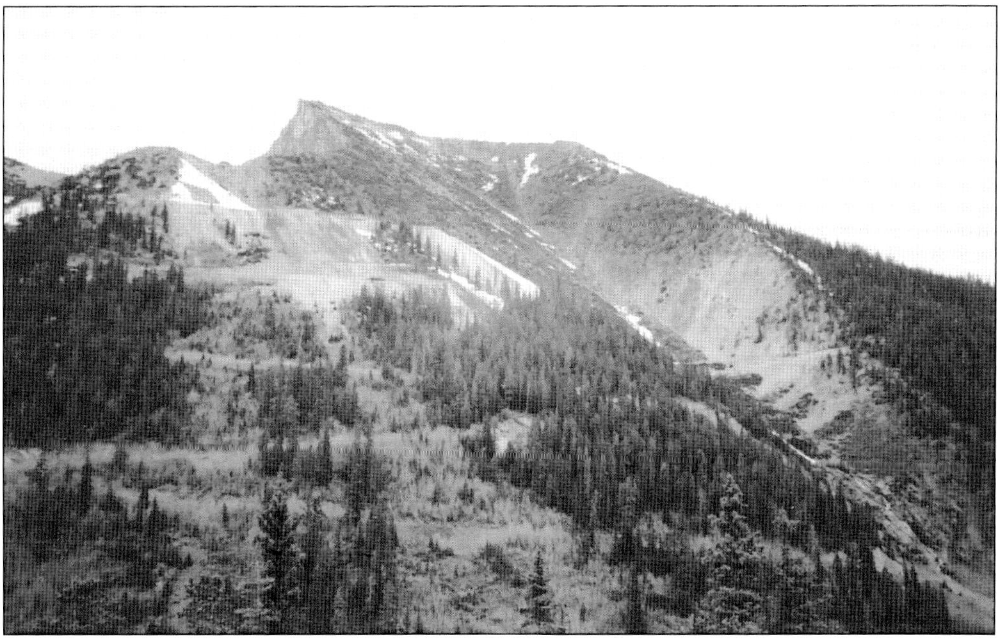

Seen here is the URAD side of Red Mountain west of Empire. The mountain is caved in on itself at the summit. The method of mining is referred to as glory hole mining. The ore is pulled out from chutes down below and loaded into dump trucks before it is dumped into the main crusher.

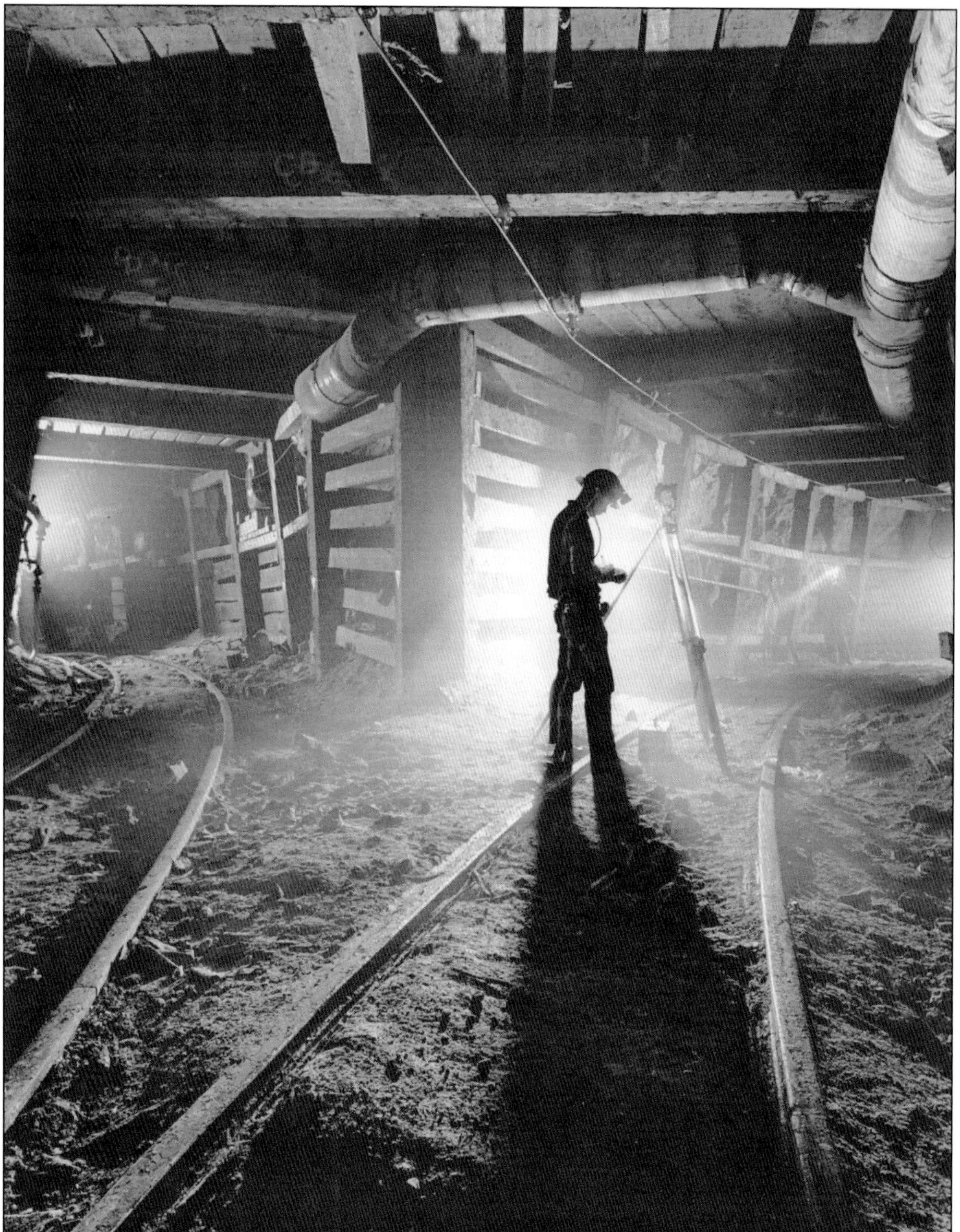

An engineer checks the proper heading for one of the new tunnels being drilled at the Climax Molybdenum Mine in 1967. There is a labyrinth of tunnels underneath Bartlett Mountain near Leadville. (CCHS.)

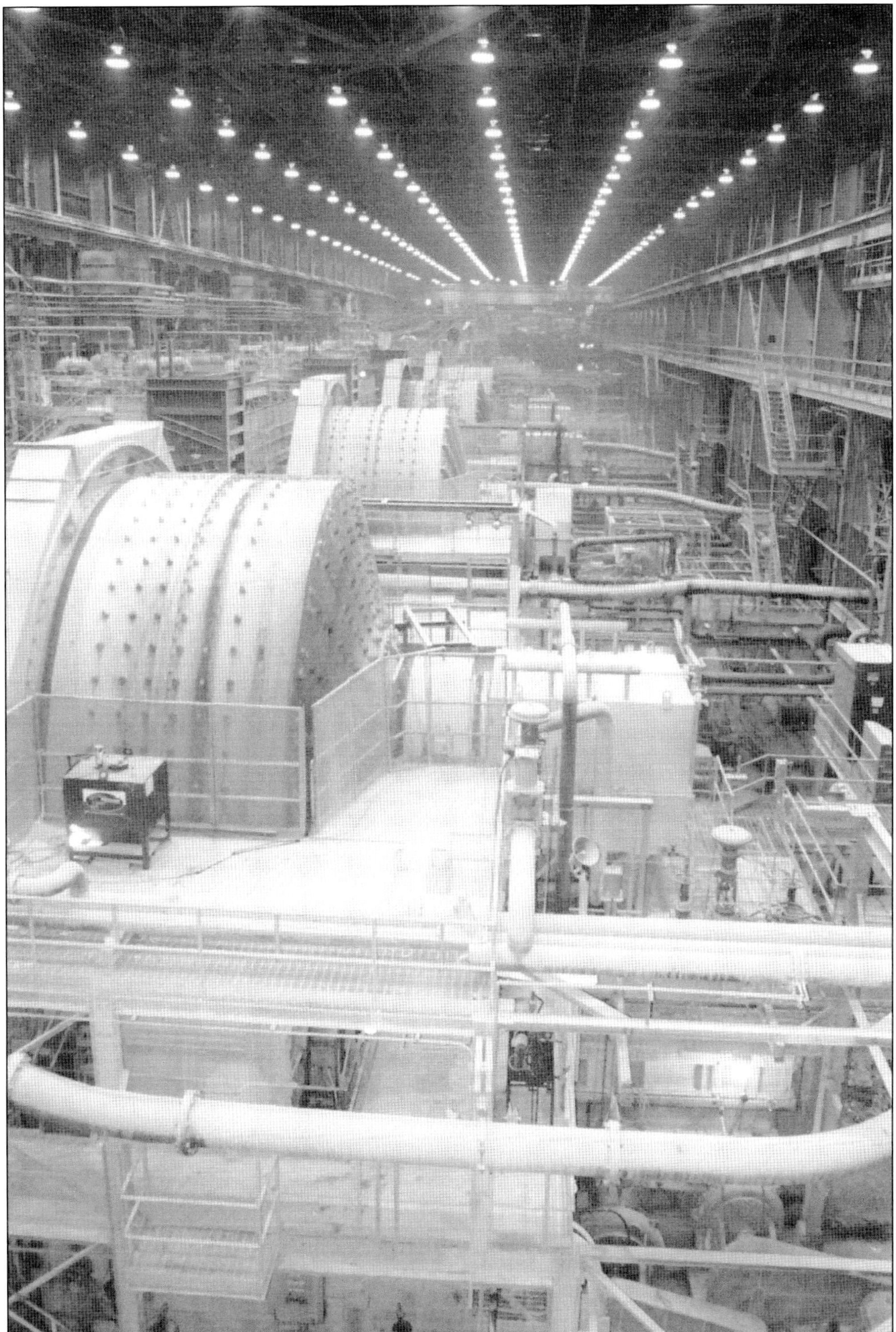

This is an interior view of the ball mills at Henderson Mill near Parshall.

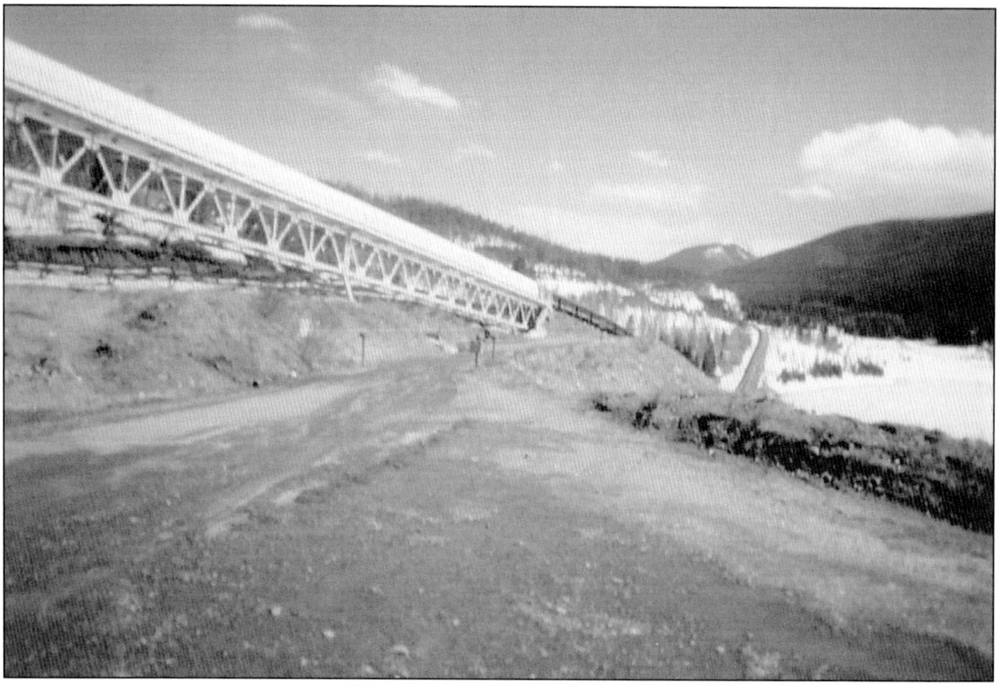

This conveyor belt was built in 2000 to replace the haulage train. As the mine went deeper, the ore was unable to be transferred to the railroad cars at the main railroad level. The conveyor belt could be adjusted down as the mine went deeper.

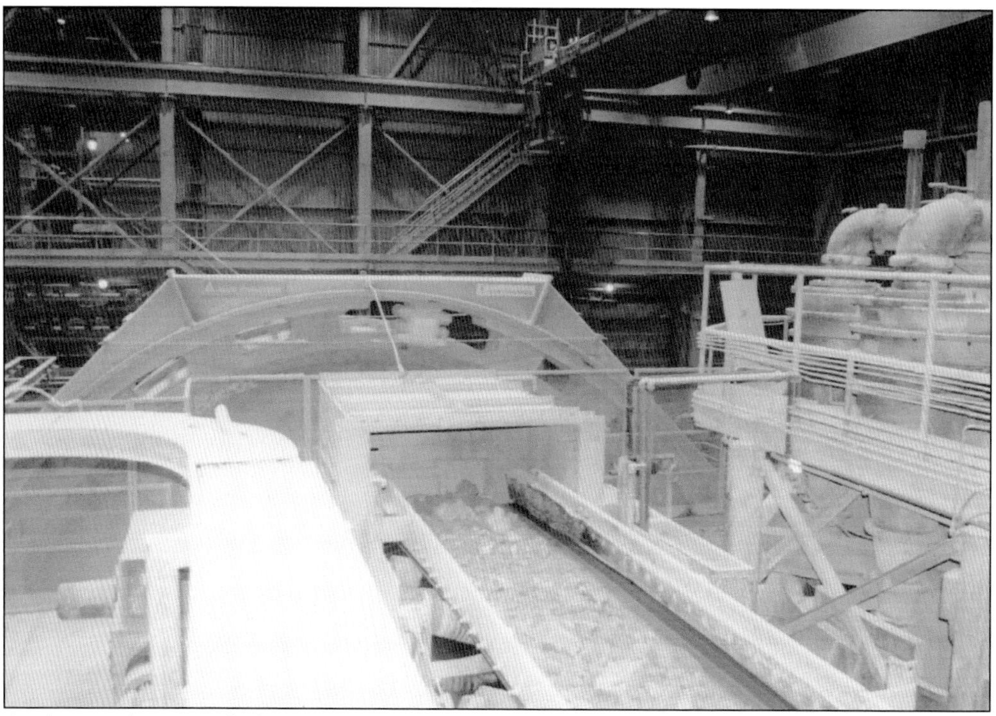

At the Henderson Mill, this is molybdenum ore going into the ball mill. A series of mills reduces the ore to a fine powder.

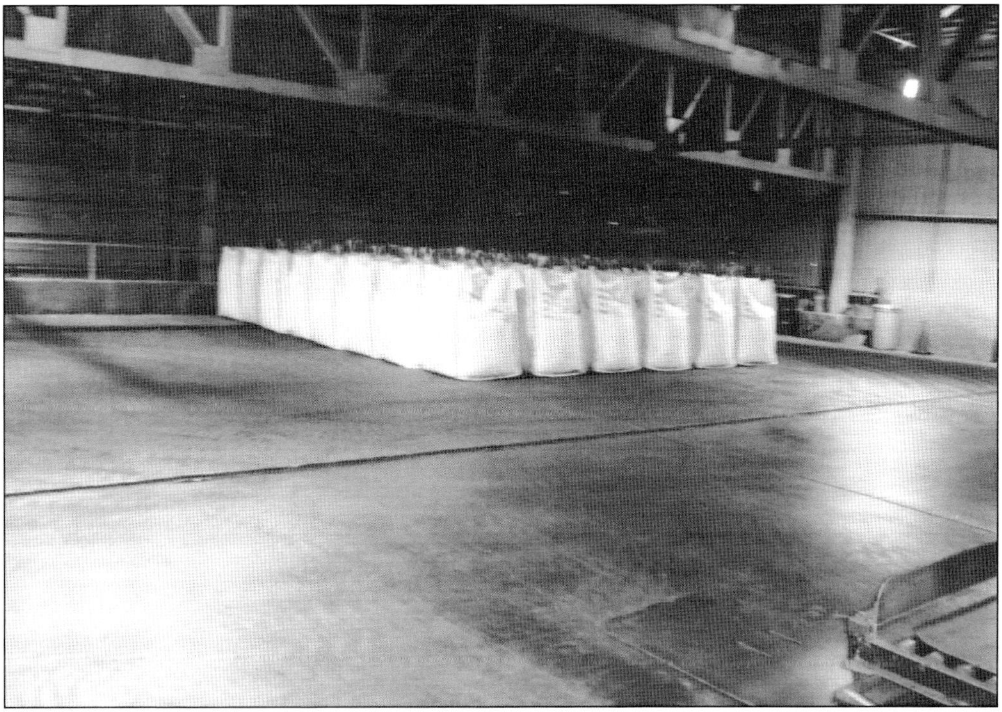

Sacks of molybdenum ore are ready for transport at Henderson Mine near Empire.

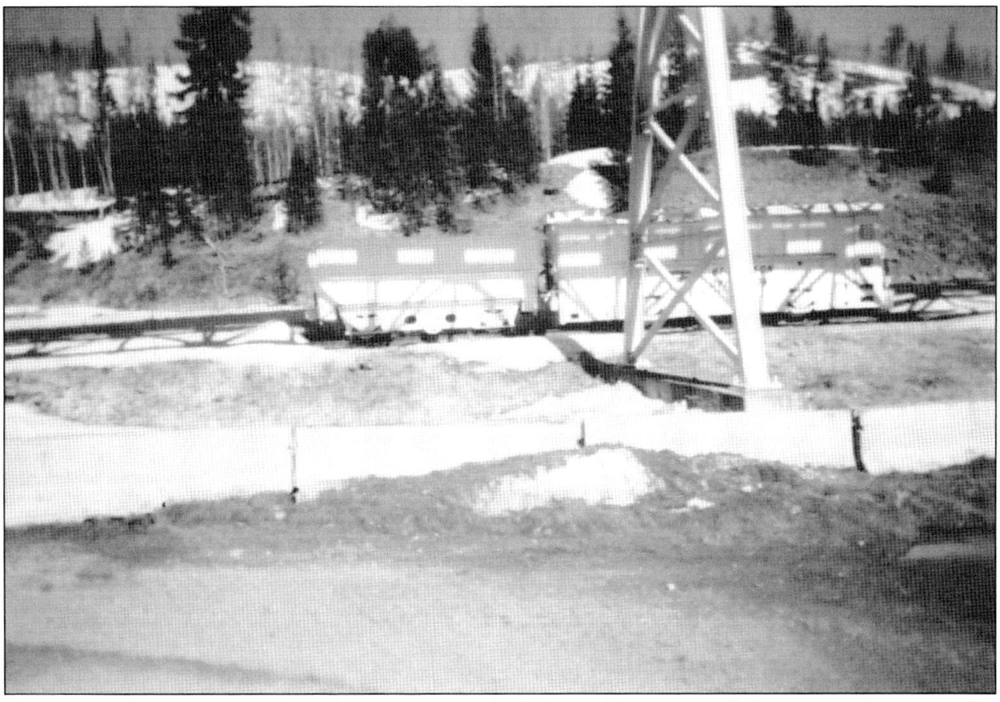

The old railcar system at Henderson Mine relied on railcars to haul the ore underneath the mountain to the mill. As the mine level grew deeper, it was difficult to get the ore to rail level, so the train was scrapped in 2000 for a conveyor belt. (CCHS.)

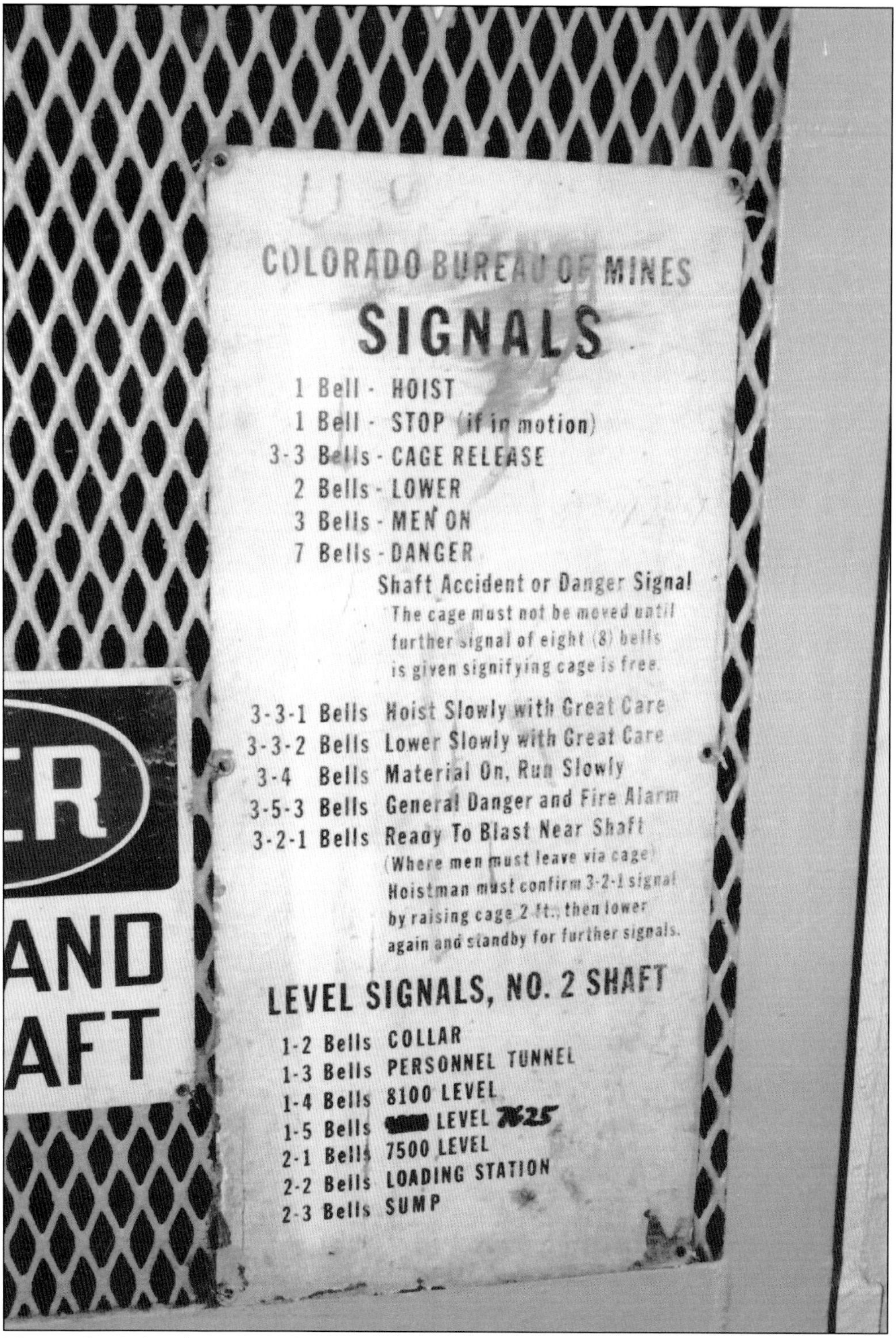

Here is the Colorado Bureau of Mines' signals sign on the elevator at Henderson Mine. Once inside the elevator, it goes down for approximately two miles.

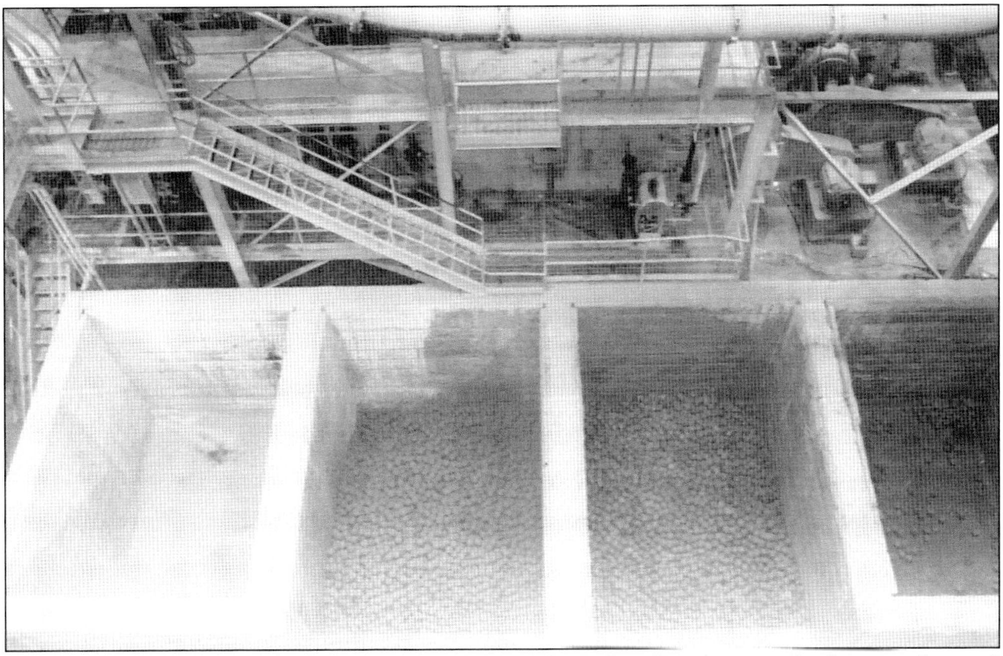

This storage bin is used for the steel balls that go into the ball mills. These used to be manufactured by the Colorado Fuel and Iron Company (CF&I) in Pueblo.

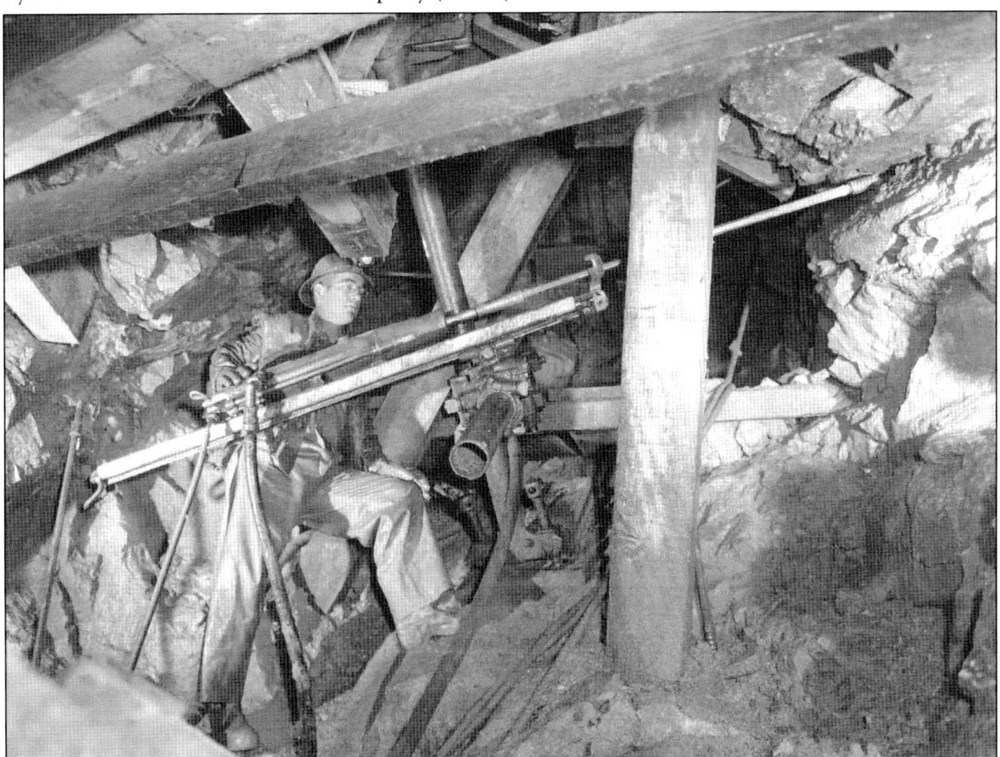

This miner uses a huge pneumatic drill to expand the tunnels below Bartlett Mountain at the Climax Mine near Leadville.

Consistent with our mission to preserve history on a local level, this book was printed in South Carolina on American-made paper and manufactured entirely in the United States. Products carrying the accredited Forest Stewardship Council (FSC) label are printed on 100 percent FSC-certified paper.

MADE IN THE USA